Religion, Evolution and Heredity

Special Issue of
*The Journal of Religious History,
Literature and Culture*
2018

Edited by
MARIUS TURDA
Oxford Brookes University

Volume 4 November 2018 Number 2
UNIVERSITY OF WALES PRESS
https://doi.org/10.16922/jrhlc.4.2

Editors
Professor William Gibson, Oxford Brookes University
Dr John Morgan-Guy, University of Wales Trinity Saint David

Assistant Editor
Dr Thomas W. Smith, University of Leeds

Reviews Editor
Dr Nicky Tsougarakis, Edge Hill University

Editorial Advisory Board
Professor David Bebbington, Stirling University
Professor Stewart J. Brown, University of Edinburgh
Dr James J. Caudle, Yale University
Dr Robert G. Ingram, Ohio University, USA
Professor Geraint Jenkins, Aberystwyth University
Dr David Ceri Jones, Aberystwyth University
Professor J. Gwynfor Jones, Cardiff University
Dr Frances Knight, University of Nottingham
Professor Kenneth E. Roxburgh, Samford University, USA
Dr Robert Pope, University of Wales Trinity Saint David
Professor Huw Pryce, Bangor University
Dr Eryn M. White, Aberystwyth University
Rt Revd and Rt Hon. Lord Williams of Oystermouth,
Magdalene College, Cambridge
Professor Jonathan Wooding, University of Sydney, Australia

Editorial Contacts
Professor William Gibson   wgibson@brookes.ac.uk
Dr John Morgan-Guy   j.morgan-guy@uwtsd.ac.uk
Dr Thomas W. Smith   T.W.Smith@leeds.ac.uk
Dr Nicky Tsougarakis   tsougarn@edgehill.ac.uk

Publishers and book reviewers with enquiries regarding reviews should contact the journal's reviews editor, Dr Nicky Tsougarakis *tsougarn@edgehill.ac.uk*

Cover illustration: adapted from the journal *Jó egészséget!* 3, 7 (July 1944). Courtesy of the Széchényi National Library, Budapest (Hungary).

CONTENTS

Acknowledgements   v
Contributors   vii

Scientific Calvinism: Eugenics as a Secular Religion
Marius Turda   1

Squaring the Circle? Two Attempts to Reconcile Darwinism
and Christianity in Late Victorian Britain
David Redvaldsen   17

From Biopolitics to Eugenics: The Encyclical *Casti Connubii*
Emmanuel Betta   39

Eugenics, Sex Reform, Religion and Anarchism in Portugal
Richard Cleminson   61

Responsible Parenthood: Reproduction and Religion in
Post-War Britain
Patrick T. Merricks   85

Index   107

ACKNOWLEDGEMENTS

This special issue on 'Religion, Evolution and Heredity' brings together papers presented at various workshops organised at Oxford Brookes University by the Centre for Medical Humanities and its Working Group on the History of Race and Eugenics, separately and in collaboration with the Oxford Centre for Methodism and Church History. One such event, organised on 8 November 2016, brought together members of two centres to discuss various aspects related to the history of medicine and religion. It is at the initiative of two participants, Professor William (Bill) Gibson and Dr John Morgan-Guy, that this special issue was commissioned for *The Journal of Religious History, Literature and Culture*. I am indebted to both of them. To Bill, in particular, I am grateful not only for the original impetus of this project but also for his guidance, advice and generosity.

I was particularly delighted that two long-time collaborators and friends, Professor Richard Cleminson and Dr Emmanuel Betta, accepted my invitation to contribute to this special issue. To work with them and with the other two contributors, Drs David Redvaldsen and Patrick T. Merricks, was a rewarding experience.

I also want to express my gratitude to Professors Christiana Payne and Joanne Begiato for their support and encouragement. Finally, my thanks to Ross Brooks and Fiona Mann for reading and commenting on two articles included here.

Marius Turda

CONTRIBUTORS

Emmanuel Betta is Associate Professor, Sapienza University of Rome. He is the author of *L'autre genèse: Histoire de la fécondation artificielle* (2017).

Richard Cleminson is Professor of Hispanic Studies, University of Leeds. He is the author of *Catholicism, Race and Empire: Eugenics in Portugal, 1900–1950* (2014).

Patrick T. Merricks is the Undergraduate Officer, Lady Margaret Hall, University of Oxford. He is the author of *Religion and Racial Progress in Twentieth-Century Britain: Bishop Barnes of Birmingham* (2017).

David Redvaldsen is Associate Lecturer in History at Oxford Brookes University. He is the author of *The Labour Party in Britain and Norway: Elections and the Pursuit of Power between the World Wars* (2011).

Marius Turda is Reader in Biomedicine and Director of the Centre for Medical Humanities, Oxford Brookes University.

SCIENTIFIC CALVINISM: EUGENICS AS A SECULAR RELIGION

Marius Turda

The outbreak of the Great War in August 1914 found William Bateson (1861-1926), the celebrated English geneticist, lecturing and attending the meetings of the British Association in Melbourne and Sydney. After one such lecture, a Scottish soldier approached him and said: 'Sir, what ye're telling us is nothing but scientific Calvinism.'[1] The encounter made an impression on Bateson, who contemplated using *Scientific Calvinism* as a title for his Australian presidential addresses, and possibly for a collection of his popular writings on genetics.[2] He never did.[3] Nor did he use 'scientific Calvinism' in connection with eugenics.[4] But there were other scientists who did. The English geneticist J. B. S. Haldane (1892-1964) was one of them. He entitled an article that he wrote in 1929 for the October issue of *Harper's Magazine* 'Scientific Calvinism', and republished it in his book *The Inequality of Man and Other Essays*, which appeared in 1932.[5] 'Will scientific Calvinism', asked Haldane, 'produce the same type of society and individual character as religious Calvinism? It is quite possible', he believed. In order for this transformation to happen, however, the eugenicists – whom Haldane described as devoting 'a large part of their energies to disapproving of their fellow-creatures' – needed to gain the public and political influence that they so eagerly sought.[6]

Although it was popular journalism, the issues discussed in this article, particularly the idea of hereditary predestination, echoed widely among the supporters of eugenics, who, almost half a century after Francis Galton (1822-1911) coined the term,[7] continued to be divided over which agency was the most important in shaping human improvement: the environment (nurture) or genetic inheritance (nature). Calvinism, as is known, promotes the idea of divine predestination. Similarly, eugenics is based on the premise that one's heredity is *given* (predestined) not *made*. Although the individual may be able to correct certain 'deficiencies' through education and self-improvement, he or she cannot escape the biological heritage bequeathed to him or her by parents and grandparents. Some hereditary legacies were more felicitous than others, eugenicists believed, but ultimately they were all *written* before the individual was born.

A year after the publication of Haldane's article, the question whether eugenics could be understood as 'scientific Calvinism' was put to three American eugenicists – Albert Edward Wiggam (1871–1957), Frederick Osborn (1889–1981) and Leon F. Whitney (1894–1973) – and their short answers were published in *Eugenics: A Journal of Race Betterment*.[8] If Wiggam charged 'the fatalistic position of the environmental position and the freedom and optimism of the theory of the hereditarian basis of behaviour', Osborn chose not to endorse either position, stating instead that the 'indefinable spiritual quality [is what] enables the individual man to make the best of his opportunities and to overcome his limitations, whether of environment or of heredity'. It was Whitney, however, who engaged more directly with the question. It was 'possible', he noted, 'to argue that eugenics [. . .] be called "scientific Calvinism"'. Calvinism meant 'that a man's spiritual fate is foreordained', while eugenics presupposed 'that a man's quality and abilities [were] determined by hereditary endowment as acted upon by environment'.[9]

In their considerations of the importance of nature and nurture, these eugenicists found neither agency sufficiently stable to allow for a final pronouncement on whether eugenics was 'scientific Calvinism'. Galton himself would have rejected 'scientific Calvinism' as a description for his theory of eugenics. The question 'whether man possess[ed] any creative power of will at all, or whether his will is also predetermined by blind forces or by intelligent agencies' was unnecessary, and he deemed the 'unending argument' about predestination as detrimental to the 'practical side of eugenics'.[10] In order for eugenics to 'be introduced into the national conscience, like a new religion' – as he put it in the paper that he read before the Sociological Society in May 1904[11] – there was need for an exploration of both morality and science, pursued simultaneously and without separating nature from nurture.[12] That is not to suggest that Galton saw eugenics as a modern secular surrogate for religion.[13] Neither am I proposing a functionalist model that defines eugenics as a secular religion simply based on the premise that its ideological content was non-Christian or anti-Christian. Not only was eugenics rarely in open conflict with religion; it did not attempt to supplant it either. To be sure, eugenics vied with organised religion over the control of reproduction and over the social and biological role of the family, but, in fact, the two had a common goal: to improve the health and morality of society. What I am offering here, moreover, is not a discussion of the relationship between religion and eugenics; others have achieved that very successfully.[14] As one scholar aptly put it, 'It is rare for studies of eugenics not to mention the question of religion. The ethical issues surrounding eugenic theories raised questions for religious believers, and this subject has featured prominently in the existing historiography'.[15] What I propose instead is to discuss eugenics as a moral philosophy concerned with the improvement of human life, with particular reference to the late nineteenth- and early twentieth-century contexts. For these reasons, this article stands askew relative to the more 'mainstream' scholarship on eugenics[16] as well as the current debates on human enhancement, which, although purporting to discuss the ethical problems surrounding the legacy of eugenics, remain in large measure reticent about it.[17]

Any scientific movement is generally regarded as hostile to religion, but I do not believe that the terms 'science' and 'religion' are so mutually exclusive.[18] In many respects, science and religion were not antithetical but complementary activities coupled in a synergetic relationship, and one upon which Galton's eugenic ideal was largely based. To be sure, eugenics teetered on religion in various ways. Importantly, eugenics (as the offspring of Darwinism and positivism) revised the traditional Platonic–Christian model of humanity's corporeality, in which the body was devalued as fallen and corrupt and ultimately dismissed as insignificant. The Christian condemnation of the body was certainly not as extreme as some nineteenth-century positivists and evolutionary scientists suggested it to be, but there is no doubt about the renewed importance that evolutionism and modern theories of heredity in general and eugenics in particular bestowed upon the body in the establishment of a new vision of humanity.[19] Just as Darwinism may be seen as challenging the hegemonic role of religion[20] and the biological fixity of the human species, eugenics may be seen as supporting the very notion of humanity as defined in terms of a hierarchy of distinct social bodies, some better biologically equipped than others. But 'eugenic qualities', as Galton was keen to emphasise, were not just 'a sound mind and body' and 'an intelligence above the average' but also 'a natural capacity and zeal for work'.[21] The healthy body and sound morals, which Galton deemed essential, needed to be accompanied by industriousness and social schemes for the betterment of men and women. In this, eugenics is as close to Christianity, with which it shares some basic moral tenets, as to Darwinism, from which it differs in some crucial aspects.[22] As was made clear by William Inge (1860–1954), later Dean of St Paul's, in the first issue of *The Eugenics Review*, the 'aim of Christian ethics is, quite definitely, the production of "the perfect man"'. And he explained further: 'The word translated perfect means full-grown, complete and entire. The perfect man is the man who has realised in himself the ideal of what a man should be.'[23] But it was not only the precepts of 'Christian ethics' that were echoed by the eugenicists. Many scholars linked the eugenic obsession with biological regeneration to the modern quest for 'a new meaning' of life and a 'new man/woman', which political ideologies such as communism, fascism and Nazism embraced so forcefully in the twentieth century.[24]

Tellingly, eugenics aimed to redefine what to be a human being was, and what were the social and biological responsibilities following from this realisation. In late Victorian Britain, Galton and others phrased the aspirations of eugenics as being the dream of a society populated by healthy individuals aware of their biological value. This vision is clearly expressed in his 1901 Huxley Lecture at the Anthropological Society. Referring to the Parable of Talents in the New Testament, Galton praised the 'good and faithful servant' of the parable, who knew how to turn his five talents into ten through wise planning. 'Whether it be in character, disposition, energy, intellect, or physical power, we each receive at our birth a definite endowment', Galton noted, 'some receiving many talents, others few; but each person being responsible for the profitable use of that which has been entrusted to him'.[25] This is why it was useful to try to consolidate those hereditary qualities, but also why great caution was necessary. Improvement, Galton believed, was to be found as much in the proper management of human aspirations as in responsible reproduction.

That Galton's eugenic ideas were hardly original needs no discussion here. The main tenets of eugenics derived equally from classical and Christian humanisms, Immanuel Kant's anthropology and the nineteenth-century scientific theories of evolution and progress – from the work of Charles Darwin, Thomas Malthus and Herbert Spencer, among others. Darwin's *The Origin of Species*, in particular, made a powerful impact on Galton, arousing in him 'a spirit of rebellion against all ancient authorities whose positive and unauthenticated statements were contradicted by modern science'.[26] Emboldened by

Darwin's work on the question of species, Galton set about defining his own interest in, as he put it, 'the central topics of Heredity and the possible improvement of the Human Race',[27] by bringing a new imaginative intensity and excitement to scientific work on the transmission of 'hereditary talent and character', to name one of his famous articles published in 1865.[28]

Galton envisioned this theory of eugenics not as an *agent* of historical change (as later eugenicists would almost feel obliged to) but as an *interpretation* of the scientific truth revealed in statistical and biometrical research. Around the end of the 1860s, he began to formulate a new vision of science devoted to heredity, which was to allow the would-be eugenicist not merely to extend his or her knowledge about biological and gender inequality (using biometrical measurements of various individuals, for instance) but to intervene actively in the processes of the human body, using the control of reproduction and through other measures. There was, too, an unrestrained endorsement of positivism as a methodology useful to the genuine improvement of the human condition. Believing in the growing acceptance of Darwinism, and thus of a new scientific doctrine about the origins of man, Galton also sought to limit the domain of organised religion in shaping human destiny, envisaging a leading role for science instead. The 'goal of Galtonian teaching', his disciple, the statistician Karl Pearson (1857–1936) noted, was 'the conversion of the Darwinian doctrine of evolution into a religious precept, a practical philosophy of life'. And he clarified this point further.

> Some may question whether we have more here than in Comte's *Religion of Humanity*. I think so, because it is freed of the ceremonialism which Comte and Gruppe demanded as a factor of religion, and it is essentially based on the acquirement of knowledge in a field of science, which had little if any existence in Comte's day.[29]

This passage reveals how Galton's interpretation of eugenics was seen early on as a scientific ontology premised upon the individual as the only one who can *accept* or *deny* his or her own improvement. Besides focusing on ways to encourage those of worthy human qualities to reproduce more regularly while discouraging those whose racial and social value was questionable, Galton's eugenics was notable for its celebration of the individual's responsibility for his or her own destiny and for the emphasis that it placed on acting on behalf of that responsibility.[30]

While organised religion stood outside his scientific agenda, Galton did not repudiate it. In fact, Galton hoped that the similarly disciplined devotion that had made religion the dominant social and cultural force for centuries would likewise inspire the future supporters of eugenics. In an article written for *The National Review* in 1894, entitled 'The part of religion in human evolution', Galton reaffirmed his conviction that 'A passionate aspiration to improve the heritable powers of man to their utmost, seems to have all the requirements needed for the furtherance of human evolution, and to suffice as the basis of a national religion, in the sense of that word as defined by J. S. Mill.'[31] I do not wish to dwell on Galton's relationship to Mill and other political economists, as others have done it more ably than I can hope to do here.[32] What I wish to point out is that, while Galton's eugenic theories seem to have been a natural consequence of his interests in heredity and evolution, the particular form that his idea of eugenics took reflected other contemporary theories as well – theories that were not situated within natural sciences but within the emerging discipline of sociology.[33]

Such a claim must be understood within its proper historical context. During the nineteenth century, the metaphysical privilege that the Enlightenment philosophers gave to man – seen as the expression of a 'universal essence' – was endorsed by the belief that scientific progress was incremental and that there was a cumulative growth of human knowledge; in other words, there was a belief in humanity's progress both culturally and biologically. Within this historical and cultural context, the ultimate objective of eugenics was defined as the creation of a functional society, subject to the rational design of modern science. Certainly, this aim was based on the recognition that religion was in 'retreat' and that secularisation was in ascendancy, particularly in the Protestant countries. Some authors, such as the Dutch theologian and politician Abraham Kuyper (1837–1920), attempted to fuse Calvinism with evolutionary sciences;[34] others, such as the secretary of the Zoological Society of London, Peter Chalmers Mitchell (1864–1945), allowed for no such synthesis. Scientists superseded priests, Mitchell declared triumphantly in 1903.[35]

In some respects, eugenics imitated religion's dream of transforming humanity. With its objectifying scientific gaze, eugenics offered the possibility of creating a society in which those individuals who were deemed less worthy were not allowed to reproduce. This gradual invasion of the private sphere of the individual has traditionally been associated with the emergence of modernity and with the birth of modern forms of systematic population control. As Michel Foucault and his followers have long argued, the birth of the modern state was also the birth of organised discipline and punishment, with the ultimate purpose of creating obedient and identical subjects.[36] Whether Galton thought in terms of purifying society from 'defective genes' or of protecting it from mixing with 'racially inferior' elements, it is clear that he believed one of the main functions of the modern state to be the achievement of the nation's racial enrichment and physical regeneration.

In light, then, of the persistently anti-religious overtones of theories of human improvement characterising the end of the nineteenth century, Galton's eugenics was bound to take a sharper meaning. It became simultaneously a biological and a social project. In the name of science and on behalf of the state, Galton fused hereditarian and cultural determinism with a modern vision of a eugenically aware society. To be sure, this view was derived also from his consistent endorsement of nature over nurture. I have discussed this aspect of Galton's theory elsewhere, so I will not repeat it here.[37] Suffice to say that Galton had highlighted the importance of nurture already, in his study on 'Hereditary Improvement', which was published in 1873.[38] In his commending of nurture, Galton dwelt on the need to create an effective 'system which shall be perfectly in accordance with the moral sense of the present time', warning that 'the ordinary doctrines of heredity', if proven to be true, aimed to 'transform the nation', biologically and morally.[39] As the history of eugenics in the twentieth century amply demonstrates, in most cases, this project also implied a new theory of modern society, guided by scientific elites, intent on controlling human reproductive patterns for social and biological purposes.

So, what did Galton's characterisation of eugenics as the 'religion of the future' really amount to? Its central message pointed to the importance of moral education. But, equally importantly, this eugenic philosophy was predicated upon a view of individual and social regeneration regulated by science. Eugenics, from this perspective, was not about encouraging the individual's public involvement, as was customary in the individualist liberal tradition, but was a means of encouraging precisely the opposite: the fulfilment of individual aspirations within the collective realm. To improve humans should become a systematic, ritualised practice, Galton recommended, and eventually eugenic harmony would be achieved after a few generations. Family history and genealogy were crucial to this eugenic narrative of human improvement. The singular focus on the history of the individual – on his or her origins and ancestry – was repeated routinely whenever the physical value of the eugenic subject was questioned and probed. The body, for Galton, was thus a heterogeneous synthesis of physicality and history in which the ephemeral biological condition of the present intersected trajectories of past inheritance. In appropriating the authority of religion for eugenics, Galton ventured to ponder what religion might be without the divine design, while, at the same time, demanding that eugenics be seen to challenge the premises underlying other scientific disciplines dealing with the human body and human relations, such as sociology and anthropology.

It is worth pausing for a moment to discuss the impact that Galton's definition of eugenics as a secular religion had upon his contemporaries. First of all, it offered the discipline of dogma, resting on the scientific authority of the natural sciences, but aiming to forge a biological theology for the future. This was a promise that eugenicists the world over were to embrace enthusiastically in subsequent decades. During the early decades of the twentieth century, there was a growing appreciation among cultural and political elites that their country's survival was wedded to its biological future, and eugenicists were the experts to supervise it.

Undoubtedly, the emergence of eugenics is intimately connected to specific historical circumstances, including empire, colonialism and racism. But, importantly, it was also linked to a general acceptance of theories of evolution by the scientific community and supported by a remarkable degree of institutional networking. Galton's first commandment – the popularisation of eugenics 'as an academic question' – served as the mantra of the First International Eugenics Congress, convened in London during July 1912. Certainly, at the time, British, American and German eugenicists were praised for their commitment to practical schemes of social and national rejuvenation derived from theories of evolution and heredity. More often than not, developments in other national contexts displayed similar features. Eugenics in France, Italy, Russia and the Scandinavian countries, for example, emerged both as a response to local conditions and as an emulation of the above-mentioned hegemonic models. In fact, this intermingling of internal and external factors dominates all national histories of eugenics.

It did not take long for eugenicists all over to embrace Galton's idea of eugenics as a new secular religion. But the problematic nature of this idea remained unresolved, allowing some eugenicists to conceive of Galton's theory as situated at the interstices of sociology and biology, while others situated it at the confluence of science and politics. In Hungary, for instance, it appealed to progressive leftist eugenicists, such as József Madzsar (1876–1940) and Zsigmond Fülöp (1882–1948).[40] In the Soviet Union, it was embraced by the biologist Nikolai Koltsov (1872–1940), who declared at the inaugural meeting of the Russian Eugenics Society in 1921:

> Eugenics has a high ideal that is also capable of giving meaning to life and moving man to sacrifices and self-limitation: to create through the conscious work of many generations an elevated type of a human, a mighty king of nature and creator of life. Eugenics is the religion of the future and it awaits its prophets.[41]

Other authors, such as the translator and writer Maximilian A Mügge (1878–?) connected Galton's eugenics to the critique of modernity and the need for a new morality, both widely popularised at the time by a host of philosophers and writers. Building on a temporal synchronicity between Galton's *Inquiries into Human Faculty* and Friedrich Nietzsche's *Thus spoke Zarathustra*, Mügge considered that the latter's concept of the 'Superman' was, in fact, the expression of the 'poetic-philosophical concept of Positive Eugenics'.[42] Mügge's was 'a potent attempt to formulate a new code of morals'[43] and other eugenicists, albeit less enamoured with Nietzsche, agreed. For the English physician Caleb Saleeby (1878–1940), eugenics was 'at once a science and a religion, based upon the laws of life, and recognising in them the foundation of society'. Saleeby believed that eugenics and religion could coexist, as they shared a desire for human perfection. 'If the struggle towards individual perfection be religious', Saleeby concluded, 'so assuredly, is the struggle, less egotistic indeed, towards racial perfection'.[44]

The eugenicist wanted to be not just a prophet, but also the priest of the new scientific religion. This is evident in 'Eugenics as a Religion', a text that the American eugenicist Charles Davenport (1866–1944) prepared as a 'sermon' for the Golden Jubilee Celebration of the Battle Creek Sanitarium held in 1916. Public gestures such as these offer insights into how some scientists conceived of the process by which eugenics was communicated to the general public in the form of a secular ritual performed to ensure the protection of the family and the race.[45]

We must be aware, however, of the fact that the religious language used by some eugenicists was as artificial as any other, and some eugenic writings performed, in this sense, an artifice. However, others, such as William Inge, used references to Christianity in a candid way, because, for him, such a reading of Galton's ideas absorbed and redeemed whatever disagreement may have existed between science and religion. 'The scientific mind', he noted in 1921, 'ought to be able to take long views, and to realise that pessimism is as little justified as optimism. [...] We are on the side of Dame Nature, and Dame Nature has a short and sharp way of punishing her rebels'.[46]

Inge transcribed the signification of Christianity (as a moral project) on to the signification of eugenics (as a biological project). For him, as for the other religious figures who embraced eugenics, the effective dissemination of eugenics and its survival as a secular religion were primarily a matter of its *form* (that is, a language that renders into biological improvement the religious ideals of sexual morality and family) and not of its *content* (that is, the replacement of organised religion by organised eugenics). In this sense, eugenics as a secular religion achieves a new meaning.

I have offered these examples (and many more can be adduced), across the political and geographical divide, of some eugenicists who reflected on Galton's programmatic text from 1904 in order to emphasise the spatial texture of the notion of eugenics as a secular religion. Surely the notion, as I have tried to suggest here, is endowed from the outset with an enormous burden: it not only documents a *transfiguration* in the ways in which individuals understand their responsibility towards future generations but offers the eugenicist the possibility of judging the work of nature itself. Undoubtedly, what operates here is a deeply embedded desire to weave into the idea of eugenics as a secular religion not only an inventory of social history but, equally importantly, a recourse to moral authority in a form that would become the emblem of some of the many attempts to rescue eugenics from its racist appropriations before and after 1945.[47]

The American sociologist John W. Slaughter (1878–1964) remarked upon this close relationship in his paper presented at the First International Moral Education Congress at the University of London between 25 and 29 September 1908. The development of a morality was central to Galton's theory of eugenics as a secular religion. According to Slaughter: 'Eugenics supplies a new moral principle. We have without doubt entered upon a new chapter in ethics based on knowledge and man's nature and conditions of his descent. This biological knowledge is demanding a corresponding sense of biological responsibility.'[48] Rooted in biology, eugenics claimed a distinct moral authority to address issues that bore directly on both individual and community, and both men and women. What emerges out of the eugenic readings of social and biological life is the need for a new morality along with a new epistemology of the human body.

As is known, in most Christian societies, it was the Church that controlled both the spiritual and the physical body of the nation. Eugenicists, in general, did not discourage religious beliefs; indeed, many of them were also practising Christians, postulating that the state's biological aims should reflect the transcendental aims of the Church. In other circumstances, the Church intervened directly in finding a solution to demands for the eugenic improvement of society, and many readers are familiar with the social and welfare activism propagated by the Catholic Church.

No one, perhaps, articulated it more clearly than Monsignor Maurice-Louis Dubourg (1878–1954), the Archbishop of Marseilles, during the conference on 'The Church and Eugenics' in 1930, organised by the Association of Christian Marriage. For Dubourg: 'If the goal of the new science [of eugenics] is, as its name indicates, to assure good offspring, it can only inspire our sympathy and find in Christian morality an auxiliary, even a very precious guide, because we profess that if God commanded man to multiply, He did not wish him to multiply poorly.'[49]

In the eyes of the Catholic Church, of course, negative eugenics was morally and religiously objectionable.[50] But, as Catholic eugenicists pointed out, Galton's version of eugenics as a secular religion was one that accepted a life dedicated to improvement and in this respect resembled the Christian conception of morality. This was made clear by the German Jesuit and biologist Hermann Muckermann (1877–1962), who, in his 1933 *Eugenik und Katholizismus* (*Eugenics and Catholicism*), insisted that Galton 'never lost sight of the need to link his science with religion'.[51] In fact, Muckermann suggested that Catholicism and eugenics were not fundamentally different as in both religion and biology certain absolute principles could be found. 'Different religious sects', he concluded, 'have different point of view, and each must respect the view of the others. But all should work together for a national eugenics on the basis of a natural ethic'.[52]

Amid the growing acceptance of abortion among Christian states in Europe and beyond, Pope Pius XI (1857–1939) issued the Encyclical on Christian Marriage, *Casti Connubii*, on 31 December 1930. The encyclical castigated the prevention of 'unworthy' life advocated by eugenicists as an expression of both excessive secularisation and the state's interference in the individual and family's private sphere.[53] In this new political staging – characterised most notably by the rise of the totalitarian regimes in Continental Europe and elsewhere, such as Latin America – something paradoxical happened that Galton could not have envisioned: eugenics eventually became a secular religion in Nazi Germany, albeit not so much by virtue of its promise of human perfection as by virtue of its transgression of, and ultimately total disregard for, human values.[54] Yet, amid growing concerns over the abuses of German racial science, biologists such as Julian Huxley (1887–1975) continued to be committed to Galton's vision of eugenics as a secular religion, with Huxley declaring in 1936: 'Once the full implications of evolutionary biology are grasped, eugenics will inevitably become part of the religion of the future, or whatever complex of sentiments may in the future take the place of organised religion.'[55] This belief in the progress of eugenics is – as I have tried to demonstrate in this article – an aspect of a much wider discussion about the role of science and religion in modern societies. Eventually, Huxley hoped,

> Science will be called on to advise what expressions of the religious impulse are intellectually permissible and socially desirable, if that impulse is to be properly integrated with other human activities and harnessed to take its share in pulling the chariot of man's destiny along the path of progress.[56]

Rightly or wrongly, many eugenicists, including Galton, felt that organised religion was cripplingly deficient in dealing with human improvement. 'Science', the French biologist Alexis Carrel (1873–1944) remarked in 1935, gave 'man the power of transforming himself. It has unveiled some of the secret mechanisms of his life. It has shown him how to alter their motion, how to mould his body and his soul on patterns born of his wishes. For the first time in history, humanity, helped by science, has become master of its destiny'.[57] This was eugenics as a modern, secular religion, based on morality and the vision of a technologically controlled humanity derived from science. Thus, while eugenics was indeed designed to remodel the individual and society, it also illustrated something else – namely the implication of transcendence, accompanying the knowledge of how the improvement of man's *nature* could in fact be achieved. Yet, for Galton, persuasion, not coercion, was eugenics' primary purpose.[58] Where later generations of eugenicists saw the commodification of state power over individual, political ruptures and racial incompatibility, Galton still pictured an underlying unity of the human race. He, like other scientists of the late nineteenth century, was aware of human imperfection, but he assumed that eugenics – together with social mobility and education – could perfect the individual. It was thus not only a claim for universalism that eugenics borrowed from religion but also the idea of perfection and salvation. This, understandably, irked some authors, including, most notably, the Austrian legal philosopher Hans Kelsen (1881–1973).[59]

What this ideal of a new eugenic religion was assumed to reveal was a profound transformation of the individual and the collective; or, in the words of the French philosopher, Emmanuel Levinas (1906–95), the 'humanity of the human'.[60] Reflecting this desire – and by relying on scientific disciplines such as sociology, biology, medicine and anthropology – eugenics rearranged the *nature* of the human body, both in *quantity* (by attempting to regulate reproduction) and in *quality* (by stimulating social and biological worth), according to a set of principles based on the laws of heredity and knowledge of the social and biological environment.

It was this portrayal of eugenics that relied on the fusion between scientific language and forms of religious and political rituals. Any attempt, therefore, to recapture how Galton and other late nineteenth-century eugenicists formulated their ideas of eugenics must inevitably contain an understanding of how ideas of evolution and heredity have competed with traditional forces, such as religion, for supremacy over the functions of the human body. When Galton spoke of eugenics as the 'new religion of the future', he hoped not only to convert coming generations to the new scientific faith but also that these new converts would establish eugenics as a universally recognised science for the improvement of the human race. Eugenics was thus articulated as a secular religion – not in its *religious sense*, however, but in its most concrete *historical sense*, as an answer to the question: How can humanity be improved? Or, in Frederick Osborn's words, eugenics should be an ideal that 'will continue to lend mystery to life and hope to our aspiration'. And then, rather mischievously, he asked: 'Is this "scientific Calvinism"?'[61]

Notes

1. See *William Bateson, F.R.S. Naturalist. His Essays & Addresses together with a Short Account of His Life* (Cambridge, 1928), p. vi. According to Guido Pontecorvo (1907-99), the Italian-born Scottish geneticist, the 'Scottish soldier' was none other than the playwright and surgeon James Bridie (1888-1951). See Wellcome Library, London, Guido Pontecorvo Papers, Correspondence between Pontecorvo and Professor Cyril Dean, Darlington, 26 November 1953, UGC198/3/1/34-1.
2. G. Radick, 'Presidential Address: Experimenting with the Scientific Past', *British Journal for the History of Science*, 49/2 (2016), 153-72.
3. He used it again in his 'Gamete and Zygote: A Lay Discourse' - the Henry Sidgwick Memorial Lecture that he delivered at Newnham College, University of Cambridge, in 1917 - to describe the 'materialistic doctrine' informing speciation in plants and animals. See *William Bateson, F.R.S. Naturalist*, pp. 201-14, esp. p. 203. See also Alan G. Cock and Donald R. Forsdyke, *Treasure Your Exceptions: The Science and Life of William Bateson* (New York, 2008), esp. pp. 448-50.
4. Bateson was often critical of eugenics, believing that it impacted negatively on the development of genetics. 'The fact is,' - he noted in 1925 in a letter to Michel S. Pease (1890-1966), another Cambridge geneticist and a founding member of the Fabian Society - 'I never feel Eugenics is my job. On and off I have definitely tried to keep clear of it. To real Genetics it is a serious - increasingly serious - nuisance diverting attention to subordinate and ephemeral issues, and giving a doubtful flavour to good materials' (quoted in *William Bateson, F.R.S. Naturalist*, p. 388).
5. J. B. S. Haldane, *The Inequality of Man and Other Essays* (Harmondsworth, Middlesex, 1937) [first published 1932], pp. 36-50.
6. Haldane, *The Inequality of Man*, p. 46. Similarly to Bateson, Haldane had an 'on and off' relationship with eugenics, but he was 'certain that it [had] a very great future as an ethical principle' (in Haldane, *The Inequality of Man*, p. 111).
7. Francis Galton, *Inquiries into Human Faculty and its Development* (London, 1883), p. 25.
8. 'Is Eugenics "Scientific Calvinism"? Is it Biological Predestination?', *Eugenics: A Journal of Race Betterment*, 3/1 (1930), 18-19.
9. 'Is Eugenics "Scientific Calvinism"?', 19.
10. F. Galton, 'Eugenics as a Factor in Religion', in F. Galton, *Essays in Eugenics* (London, 1909), p. 69.
11. F. Galton, 'Eugenics: Its Definition, Scope and Aims', in F. Galton, *Essays in Eugenics*, p. 42.
12. F. Galton, *English Men of Science: Their Nature and Nurture* (London, 1874). See also R. Schwartz Cowan, 'Nature and Nurture: The Interplay of Biology and Politics in the Work of Francis Galton', in *Studies in the History of Biology*, 1 (1977), 133-208.
13. In the way that the philosopher of science defines evolution as a secular religion. See Michael Ruse, 'Is Evolution a Secular Religion?', *Science*, 299/5612 (2003), 1523-4.
14. For the North American context, see C. Rosen, *Preaching Eugenics: Religious Leaders and the American Eugenics Movement* (New York, 2004); S. M. Leon, *An Image of God: The Catholic Struggle with Eugenics* (Chicago, 2013); and D. L. Durst, *Eugenics and Protestant Social Reform: Hereditary Science and Religion in America, 1860-1940* (Eugene, OR, 2017). For the British context, see P. T. Merricks, *Religion and Racial Progress in Twentieth-Century Britain: Bishop Barnes of Birmingham* (Basingstoke, 2017); for the European context, see M. Turda and A. Gillette, *Latin Eugenics in Comparative Perspective* (London, 2014).

[15] G. J. Baker, 'Christianity and Eugenics: The Place of Religion in the British Eugenics Education and Society and the American Eugenics Society, c. 1907–40', *Social History of Medicine*, 27/2 (2014), 281.

[16] For example, Debbie Challis, *The Archaeology of Race The Eugenic Ideas of Francis Galton and Flinders Petrie* (London, 2013).

[17] There is now a vast literature on the topic. See, for example, M. J. Selgelid, 'Moderate Eugenics and Human Enhancement', *Medicine, Health Care and Philosophy*, 17/1 (2014), 3–12. See also N. Agar, *Liberal Eugenics: In Defence of Human Enhancement* (Malden, MA, 2004). For a different approach, see M. Ekberg, 'Eugenics: Past, Present, and Future', in M. Turda (ed.), *Crafting Humans: From Genesis and Eugenics, and Beyond* (Goettingen, 2013), pp. 89–108.

[18] On this point I am much indebted to P. J. Bowler, *Reconciling Science and Religion: The Debate in Early-Twentieth-Century Britain* (Chicago, 2001).

[19] R. J. Halliday, 'Biology and Politics: Some Victorian Perspectives', *Journal of Social and Biological Structures*, 2/5 (1979), 119–31; and David Redvaldsen, 'Squaring the Circle? Two Attempts to Reconcile Darwinism and Christianity in Late Victorian Britain' in this issue.

[20] Michael Ruse argues that Darwinism did in fact become 'a secular religion, in opposition to Christianity. In the second half of the nineteenth century and into the first part of the twentieth century Darwinian evolutionary thinking [. . .] became a belief system countering and substituting for the Christian religion: a new paradigm' (in M. Ruse, *Darwinism as Religion. What Literature Tells us about Evolution* (New York, 2016), p. 82).

[21] Sir F. Galton, 'Eugenic Qualities of Primary Importance', *The Eugenics Review*, 1/2 (1909), 76.

[22] See D. B. Paul, 'Darwin, Social Darwinism and Eugenics', in J. Hodge and G. Radick (eds), *The Cambridge Companion to Darwin*, 2nd edn (Cambridge, 2009), pp. 219–45.

[23] W. R. Inge, 'Some Moral Aspects of Eugenics', *The Eugenics Review*, 1/1 (1909), 33.

[24] As discussed at length by R. Griffin in his *Modernism and Fascism: The Sense of a Beginning under Mussolini and Hitler* (Basingstoke, 2007).

[25] F. Galton, 'The Possible Improvement of the Human Breed under the Existing Conditions of Law and Sentiment', in Galton, *Essays in Eugenics*, p. 3.

[26] F. Galton, *Memories of My Life* (New York, 1909), p. 287.

[27] Galton, *Memories of My Life*, p. 288.

[28] F. Galton, 'Hereditary Character and Talent', *Macmillan's Magazine*, 12/68 and 70 (1865), 157–66 and 318–27.

[29] K. Pearson, *The Life, Letters and Labours of Francis Galton*, vol 3a (Cambridge, 1930), p. 93.

[30] This is what the philosopher F. C. S. Schiller (1864–936) defined as the 'moral ideal' of eugenics. See F. C. S. Schiller, 'Eugenics as a Moral Ideal: The Beginning of a Progressive Reform', *The Eugenics Society*, 22/2 (1930), 103–9.

[31] Pearson, *The Life, Letters and Labours of Francis Galton*, vol 3a, pp. 92–3.

[32] See, for example, Chris Renwick, 'Eugenics, Population Research, and Social Mobility Studies in Early and Mid-Twentieth-Century Britain', *The Historical Journal*, 59/23 (2016), 845–67.

[33] R. J. Halliday, 'The Sociological Movement, the Sociological Society and the Genesis of Academic Sociology in Britain', *The Sociological Review*, 16/3 (1968), 377–98.

[34] A. C. Flipse, 'Against the Science–Religion Conflict: The Genesis of a Calvinist Science Faculty in the Netherlands in the Early Twentieth Century', *Annals of Science*, 65/3 (2008), 363–91.

[35] P. C. Mitchell, preface to E. Metchnikoff, *The Nature of Man: Studies in Optimistic Philosophy* (London, 1903), p. ix.

36. C. Hanson, 'Biopolitics, Biological Racism and Eugenics', in S. Morton and S. Bygrave (eds), *Foucault in an Age of Terror* (London, 2008), pp. 106–17.
37. M. Turda, 'Race, Science, and Eugenics in the Twentieth Century', in A. Bashford and P. Levine (eds), *The Oxford Handbook of the History of Eugenics* (New York, 2010), pp. 62–79.
38. F. Galton, 'Hereditary Improvement', *Fraser's Magazine*, 7/37 (1873), 116–30.
39. Galton, 'Hereditary Improvement', 116.
40. See M. Turda, *Eugenics and Nation in Early 20th Century Hungary* (Basingstoke, 2014), pp. 53–6.
41. N. K. Koltsov, 'Improvement of the Human Race' [1921], in V. V. Babkov, *The Darwin of Human Eugenics*, trans. from the Russian by Victor Fet (Cold Spring Harbor, NY, 2013), p. 86.
42. M. A. Mügge, 'Eugenics and the Superman: A Racial Science, and a Racial Religion', *The Eugenics Review*, 1/3 (1909), 185.
43. D. Stone, *Breeding Superman: Nietzsche, Race and Eugenics in Edwardian and Interwar Britain* (Liverpool, 2002), p. 62.
44. C. W. Saleeby, *Parenthood and Race Culture: An Outline of Eugenics* (London, 1909), pp. ix and 304.
45. Rosen, *Preaching Eugenics*, pp. 92–4.
46. W. R. Inge, 'Eugenics and Religion', *The Eugenics Review*, 12/4 (1921), 265.
47. See F. Cassata, *Eugenetika sensa tabù. Usi e abusi di un concetto* (Turin, 2015).
48. J. W. Slaughter, 'Eugenics and Moral Education', in G. Spiller (ed.), *Papers on Moral Education communicated to The First International Moral Education Congress* (London, 1908), p. 381.
49. Quoted in W. H. Schneider, 'The Eugenics Movement in France, 1890–1940', in M. B. Adams (ed.), *The Wellborn Science: Eugenics in Germany, France, Brazil, and Russia* (New York, 1990), p. 80.
50. See I. Richter, *Katholizismus und Eugenik in der Weimarer Republik und im Dritten Reich. Zwischen Sittlichkeitsreform und Rassenhygiene* (Paderborn, 2001).
51. H. Muckermann, 'Eugenics and Catholicism' [1933], in P. M. H. Mazumdar (ed.), *The Eugenics Movement: An International Perspective*, vol 4 (London and New York, 2007), p. 21.
52. Muckermann, 'Eugenics and Catholicism', p. 67.
53. E. Betta, 'From Biopolitics to Eugenics: The Encyclical *Casti Connubii*' in this issue.
54. See Turda and Gillette, *Latin Eugenics in Comparative Perspective*, pp. 103–28.
55. J. S. Huxley, 'Eugenics and Society', *The Eugenics Review*, 28/1 (1936), 11. See also Chris Renwick, 'New Bottles for New Wine: Julian Huxley, Biology and Sociology in Britain', *The Sociological Review Monographs*, 64/1 (2016), 151–67.
56. J. Huxley, 'Religion as an Objective Problem', in J. Huxley, *Man in the Modern World* (London, 1940), p. 141.
57. A. Carrel, *Man, the Unknown* (New York, 1935), p. 241.
58. C. P. Blacker, *Eugenics: Galton and After* (London, 1952), pp. 103–23.
59. See Hans Kelsen, *Secular Religion: A Polemic Against the Misinterpretation of Modern Social Philosophy, Science and Politics as 'New Religions'* (Vienna and New York, 2011).
60. E. Levinas, *Humanism of the Other*, trans. by Nidra Poller (Champaign, IL, 2006) [first published in 1972].
61. 'Is Eugenics "Scientific Calvinism"?', 19.

SQUARING THE CIRCLE? TWO ATTEMPTS TO RECONCILE DARWINISM AND CHRISTIANITY IN LATE VICTORIAN BRITAIN

David Redvaldsen

In 1859, Charles Darwin (1809–82) published a monumental book in the history of science: *On the Origin of Species by Means of Natural Selection*. Although the work unleashed shock waves, the idea of evolution, species changing gradually over time, was not completely new. The French naturalist Jean-Baptiste Lamarck (1744–1829) and Darwin's grandfather, Erasmus Darwin (1731–1802), had not believed in the fixity of the species either. In 1858, Darwin, an independent scientist, had been engaged in writing a lengthy tome on evolution when he received a letter from fellow naturalist Alfred Russel Wallace (1823–1913) asking for advice on publishing a paper containing the self-same findings that Darwin had been developing for a number of years. To establish provenance for the theory, Darwin quickly began writing a shorter version consisting of the main ideas only. The resulting book became a landmark in tracing the origins of ourselves and shattered the complacent and self-satisfied world view of many Victorians.

Darwin had taken care to offend Christians as little as possible. An alternative title for the book, *The Preservation of Favoured Races in the Struggle for Life*, implied that there was an element of design behind the evolutionary process. In the second edition of the book he inserted a few more references to the Creator and mentioned that parthenogenesis (virgin birth) was not well understood, in an attempt to mollify believers.[1] Nevertheless, many Christians were severely perturbed by the work's implications for the biblical account of Creation and its unspoken challenge to their faith. Emotionally, they were outraged by the dichotomy between 'human' and 'animal' being found to be relative rather than absolute. Darwin did not state categorically that humans shared ancestors with simians until in *The Descent of Man and Selection in Relation to Sex*, published in 1871, but the popular understanding of the first book was that we all descend from ape-like ancestors. This added to the contemporary realisation of naturalists, buoyed by discoveries of geological time and fossil records showing associated sequences of animal populations, that the Earth was millions of years old, not the few thousand years envisaged in the Bible.[2] This made it very difficult to hold on to traditionalist religious teachings. Christian doctrine seemed ripe for reconsideration in light of Darwinism and modern science.

The account in Genesis of the world being created in seven days and all the species currently in existence fashioned on the third, fifth and sixth days had to be regarded as allegorical. Since the modification of a species happened gradually, over an extended period of time, and to a large number of individuals, logically, there could not be a unique ancestor couple of modern humans. Unlike what the Bible claimed, early humans had certainly not had lifespans much longer than contemporary people, the various species had not all been herbivores until after the Flood and there had always been a struggle to survive. In fact, it seemed doubtful that there had ever been a Fall and the concept of sin had to be viewed alongside the most well-adjusted individuals surviving longer and leaving more progeny. Therefore Darwinism did not just dispense with biblical myths, but went to the heart of the doctrine.

Christianity taught that humans possessed immortal souls, which, after the death of the bodies that encased them, would be judged and continue their existence in either Heaven or Hell. No other species possessed souls, so their lives would therefore cease at bodily death. Since there had been no separate creation of human beings and there was no definite point in evolution at which one type of simian let go of its animal nature and became a human, how the soul had been implanted seemed a mystery. A situation might arise in which some members of recognisably the same species had souls and others had not. But God's mercy for humans with souls was shown to be highly limited, too, as he allowed large swathes of them to die in order to continue modifying the species. Was God in fact a social Darwinist? Why should he be so heartless to his creation when there probably had never been a Fall? And why should Christ, who was the perfect man, have given his life so that those inferior to him could live when the law of nature, instituted by God, was the opposite: the inferior must suffer and die in order for the best to be able to survive and multiply?

It is clear that the religious ramifications of Darwinism raised some very serious objections to Christianity. It went beyond the caricature of the famous debate at the Oxford Museum in June 1860, when the Bishop of Oxford, Samuel Wilberforce, locked horns with the Darwinian scientist Thomas Henry Huxley and asked a rhetorical question about which of Huxley's immediate ancestors were apes.[3] In this article we will look at two Christian responses to Darwinism in Victorian Britain, one from a self-proclaimed Catholic and one from a Protestant.

Aubrey Moore and Henry Drummond in context

The Anglo-Catholic Aubrey Moore (1848–90) was an Oxford theologian who was Fellow of St John's College (1872–6) and became Dean of Divinity of Magdalen College a few weeks before his untimely death. Henry Drummond (1851–97) also died in his forties. He was a Scottish professor of science, and a preacher. Both belonged to learned environments that might be capable of mustering the necessary intellectual rigour to defend Christianity. Oxford is the oldest university in the British Isles and its university church has heard sermons from many leading thinkers, including John Wesley (1703–91) and John Henry Newman (1801–90), who was vicar there from 1828 until 1843. Scottish Protestantism might also hold answers to some of these intellectual challenges, especially because it is beholden to John Knox (c.1514–72), who believed in double predestination (election and reprobation) – a theological equivalent to social Darwinism.

Science had a somewhat ambiguous status at Oxford. When the honour school of natural science was established in 1850, critics had feared that it would impinge on the university's 'true purposes' – namely providing the sons of gentlemen with a liberal arts education.[4] The above-mentioned museum – a temple to science – was only narrowly approved in December 1854, by a vote of 70 ayes to 64 nays.[5] But the university did appoint professors in the natural sciences, some of whom welcomed Darwin's theory of evolution, including the chemist and botanist Charles Daubeny (1795–1867) and the comparative anatomist George Rolleston (1829–81).[6] A number of clergymen were also willing to accommodate evolution. The day after the clash between Wilberforce and Huxley, the Reverend Frederick Temple (1821–1902), headmaster of Rugby School, preached a sermon to members of the university and outside scientists gathered in Oxford, stating that God's creation happened through the laws that scientists discover.[7] The most important influence on Aubrey Moore's development, however, was not the politics of science at his alma mater, but the network he belonged to while there. Moore belonged to the younger generation of the Oxford Movement, epitomised by the 'Holy Party', which began meeting in the early 1870s.[8] This reading group consisted of committed Christians who wished to bridge the gap between modern thought and their Anglo-Catholic theology. Among its members we find Edward Stuart Talbot (1844–1934), who in due course became Bishop of Lincoln; Charles Gore (1853–1932), a future Bishop of Oxford; and Henry Scott Holland (1847–1918), a canon of St Paul's and an influential Christian socialist.[9] They all revered the greatest contemporary Oxford philosopher: the idealist Thomas Hill Green (1836–82).

The Scottish Free Church, to which Drummond belonged, was also riven by debates on the role of science and the acceptability of evolution. There was a geographic cleavage, in that its Gaelic-speaking Highlander faction was suspicious of science and all modern thought that might contradict the Bible. Conversely, the leader of the Church in the 1830s and 1840s, Thomas Chalmers (1780–1847), was pleased to attend an early meeting of the British Association for the Advancement of Science that took place in Cambridge in 1833. He had scientific associates who were committed to evangelical Christianity and saw the pursuit of mysteries in the natural world as an Enlightenment endeavour. These included Sir David Brewster (1781–1868), a physicist who became the Principal of the University of Edinburgh in 1854; John Fleming (1785–1857), Professor of Natural History at Aberdeen; and Hugh Miller (1802–56), a geologist who had once been a stonemason.[10] So enraptured were they with science that they described their religion as 'experimental'. But Brewster was highly sceptical about evolution, as was the minister John Duns (1820–1909), who taught natural science at the Free Church's New College in Edinburgh after 1864.[11] In echoes of what had occurred in Oxford, Huxley's visit to Edinburgh in January 1862, when he declared that humans were of the same lineage as apes and attacked the biblical account of origins in Genesis, had raised hackles among churchgoers.[12]

The power of religion waned over the course of Queen Victoria's reign, concomitant with the march of modernity. When Robert Rainy (1826–1906), the next dominant Free Church leader after Chalmers, became Principal of New College in Edinburgh in 1874, he chose the title 'Evolution and Theology' for his inaugural lecture.[13] In it he largely accepted Darwinism and claimed that its scientific validity was unproblematic for theology. What distinguished humans from other species anyway was not their frames but their minds. Similarly, those who saw religious authority as paramount were in retreat at the University of Oxford. In 1871, Parliament repealed the Test Act relating to universities, meaning that non-Anglicans could now receive degrees at Oxford. After the passage of the Universities of Oxford and Cambridge Act of 1877, the university commissioners began reducing the number of fellowships held by clergymen.[14] In 1882, the commissioners drew up a statute setting the minimum number of clerical fellows at each college at just one.[15]

The two apologists under consideration here differed considerably in their educational backgrounds and how they engaged with science. There was nothing scientific in Aubrey Moore's conventional route towards becoming an Oxford don. The son of a clergyman and religious author, he won a scholarship to St Paul's School in London and from there went up to Oxford, where he gained first-class honours in classics, graduating in 1871. In 1880 he began tutoring at Keble College and became friendly with two science lecturers there: Edward Bagnall Poulton (1856–1943) and Sir John Conroy (1845–1900).[16] It was Poulton who introduced Moore to recent debates involving evolution, and in 1883 Conroy recruited Moore as a speaker at a seminar devoted to the relationship between natural science and Christianity.

Henry Drummond had a far stronger scientific background. He was born and raised in Stirling, where his uncle had helped set up an agricultural museum.[17] In 1866 he matriculated at the University of Edinburgh, where he studied the arts but left without taking a degree. From 1870 to 1873, however, he combined undertaking the divinity course at New College, Edinburgh with pursuing part-time studies in geology, botany, zoology and chemistry at the university.[18] In 1877 he became a lecturer in natural science at another of the Free Church's seminaries: New College, Glasgow. In 1879 he joined a geological expedition to the Rocky Mountains in the United States, and the next year he was elected to a Fellowship of the Royal Society of Edinburgh. One of his sponsors on that occasion was Sir William Thomson (1824–1907), who would later become world renowned as Lord Kelvin.[19] Both Moore and Drummond were ordained ministers, but Drummond had also been involved in lay preaching as he became an assistant to the American evangelists Dwight Moody (1837–99) and Ira Sankey (1840–1908) in 1874, travelling with them to various English and Irish cities.[20]

While the historian David Bebbington felt that Moore had defended Christianity more successfully than Drummond, the latter is far better known, and his writings are characterised by greater originality.[21] We will give in-depth treatment to Drummond's two best-known books before, towards the end of this article, reverting to Moore's Christian apologetic essays on evolution.

Darwin answers his theist detractors

Preceding either of these churchmen, however, was Darwin himself, whose second book, *The Descent of Man*, answered his critics, whether they were arguing from a scientific or a theological standpoint. Darwin was among those who had lost their religious faith on the basis of the scientific findings that had become known in his lifetime. Originally, he had thought that natural selection and gradual change over time were entirely compatible with divine creation. He had sent an advance copy of *The Origin of Species* to Charles Kingsley (1819–75), the rector of Eversley in Hampshire, and soon to be the author of *Water Babies* (1863), a classic children's novel. Kingsley had written back to say that, from his point of view, it seemed just as noble for the Creator to make primal forms capable of self-development as having to intervene periodically in natural history to sort out the emerging gaps.[22] So impressed was Darwin with the response that he included these lines in the second edition of his book. His first position in the controversy was that he was merely extending natural theology, epitomised by William Paley's *Natural Theology; or, Evidences of the Existence and Attributes of the Deity* (1802), which he knew intimately as an undergraduate at Cambridge.

Later, in the 1860s, Darwin accepted that evolution was not exactly conducive to the idea of a wise and beneficent Creator guiding the process. The theological problem of suffering had existed long before Darwin, but it received new embers to its flame through the theory of evolution. The scale of suffering inherent in the animal kingdom brought about by natural selection was as vast as the process of evolution was lengthy. And what meaning had been served by species that were now extinct? Had their suffering been without purpose?[23] In order to minimise the scale of opposition to his theory, Darwin tried to show how theism was not necessarily shaken by evolution. He never became an outright atheist, preferring to call himself an agnostic, and he also nurtured some sympathy for deism, as he thought a creation implied a Creator, working through natural laws. *The Descent of Man* was a longer book that probably contained much of the material that he had intended to publish in the abandoned work in which he intended to give the theory to the world. Writing more confidently after evolution had received support, he dealt with the objections of many theists and, in so doing, he naturally moved further away from faith.

To religious people, it was vital to show that humans were special and in a completely different category to animals. Otherwise they could not maintain their preferred position that God cared for humankind but had little regard for animals. For Christians, the chasm between humans and animals was that only the former had souls, and therefore many argued that evolution could not be true as it presupposed the common origin of humans and simians – animals belonging to the ape family. Their method was to point to the very real differences in aptitudes between our species and various animals. In response, Darwin would lay out facts showing, often anecdotally, how the chasm was not quite as wide as his detractors imagined. Humans were generally more intelligent than animals, but there was an overlap between them. In the zoological gardens of Regent's Park in London were a baboon and a New World monkey, both of which were noted for their high intelligence.[24] Conversely, idiots often shared animal characteristics such as carefully smelling every morsel of food before eating.[25] Animals varied in abilities in just the same way as humans do. Humans and animals shared many instincts, among which were self-preservation, sexual love, maternal love for newborn offspring, and the sucking mechanism in infants.[26] Animals also showed more complex emotions, shared by humans, such as jealousy in the case of dogs and monkeys, which proved a desire to be loved as well as feeling love themselves.[27] Dogs, cats, horses and even birds have vivid dreams, which shows that they have the power of imagination.[28] Given an imagination and memory, how can we be certain that an old dog does not reflect on or recollect its past life?[29] That would be a form of self-consciousness. Darwin thus dismissed the idea that there was a complete dichotomy between humans and animals beyond, obviously, belonging to different species.

Theists would also try to establish the limitations of natural selection by pointing to humans occasionally sacrificing themselves for others. Darwin argued that this was derived from the social instinct of herding together in order to aid each individual's chances of survival. The first moral codes had prohibited anything that tended to decrease the tribe's power. Showing regard for fellow members of one's tribe was encouraged by common endeavours, which then became a habit. Other theists would point to godliness and moral standards being innate to man. Darwin thought that believing in God was ennobling but that it certainly was not universal across the world. He said that belief in evil spirits was more prevalent than belief in a beneficent Deity.[30] As for a minimum of moral standards, no such generally agreed morality existed or had ever been shown to exist. For instance, Darwin considered slavery a great evil, yet it had only recently been abolished in civilised nations.[31] Hospitality and kindness to strangers were not considered virtues by most savages; they were instead indifferent to the suffering of people not belonging to their tribe or even, said Darwin, delighted in witnessing these. God or innate ideas of ethics were simply not a part of the equation.

Darwin's challenge to theism was not trivial. It is a fulcrum of atheism today. The only obvious remaining avenue to vindicate faith intellectually might be to attack Darwin's use of facts. But because he had spent so long in formulating his theory, it was richly furnished with examples from the work of others, in even the smallest of details. Far from skirting over difficulties, he freely identified them and answered potential critics in a whole chapter of *The Origins of Species*. Theists therefore had a choice between either coming to terms with Darwinism or becoming a counterculture outside the mainstream, in which different beliefs and norms applied. Drummond chose a thoroughgoing acceptance of Darwinism, as he was in any case fascinated with science. Then he tried to show that the social habits and instincts that Darwin had identified were God-given. What was so attractive about this was that people of faith could maintain their theism while fully accepting the latest scientific discoveries.

Henry Drummond's reconciliation of science and religion

Drummond's career was characterised by huge success. In June 1883, his book *Natural Law in the Spiritual World* was published, based on articles he had written in a journal called *Clerical World* in 1881 and 1882.[32] Helped by a complimentary review in *The Spectator*, within a year it was selling 1,000 copies a month.[33] Drummond was made a professor in 1884. In 1893, he was invited to give the Lowell Lectures at the institute of the same name in Boston, Massachusetts. Tickets for this lecture course were so sought after that a ring of speculators made a heady profit selling them on the black market.[34] The Lowell Lectures became his next book, *The Ascent of Man*, published in 1894. The secret of his success was perhaps the combination of teaching science at an academic institution with spreading the gospel to workers on Sundays through the Possilpark mission. In a world that was becoming increasingly sceptical, he showed, through his example, that it was possible to embrace the latest developments in science while holding on to faith. As he said in 1892, 'Let science and religion go each in its own path, they will not distort each other. The contest is dying out.'[35] People were no longer expecting the Bible to provide scientific facts, which had been an anachronism. His lack of dogmatism appealed to many. For instance, in 1893 he told students at Harvard University that it was possible to follow Christ until a better leader and teacher came along.[36] By making that undertaking, one became a Christian; it was no harder than that. The Darwinian revolution in the sciences and evangelism, particularly being involved with Moody and Sankey, formed the backdrop to Drummond's life.[37]

It is clear that Drummond's approach and ideas struck a chord. But how well did he argue, and to what extent was he actually able to reconcile religion and science? *Natural Law in the Spiritual World* contained chapters on themes such as 'biogenesis', 'degeneration' and 'parasitism', endeavouring to bring out the spiritual aspect of these. The real attempt to make the argument that the laws of the spiritual world were the same as those in nature came only in the introduction. Drummond's ideas paid heed to science, but no more convincingly than what the novelist H. G. Wells was to do a decade later, in his novel *The Time Machine* (1895), in explaining how time travel might be possible.[38] For a start, it is taken for granted that such a thing as the spiritual world exists. The analogy with the political world of Walter Bagehot or the social world of Herbert Spencer does not quite work, because we know for sure that there is a society and that it is governed by politics. Drummond claimed that these authors had brought natural law into their concerns and that he was merely following their example.[39] Using an observation by Spencer, Drummond gives a reason for eternal life: an organism in a perfect, non-changing environment would not need to adapt itself as there would already be complete harmony between it and its surroundings.[40] Drummond adds that to know God is to correspond with such an environment. This seems to contradict the principle of growth or ageing, which Drummond elsewhere claims is a spiritual law, too. Christianity 'generates the necessary condition for carrying on the organism successfully, from stage to stage'.[41]

To the Christian, the spiritual is dominant over the material. For Drummond, the material or organic world is a scaffolding leading up to the spiritual world. Once all have passed through the organic world, the scaffolding will be taken down: in other words, the Earth will come to an end as foretold by Christ and in Revelation. Each of these two worlds follows the same laws, though the organic world is described by science and the spiritual world by scripture. Entry to each requires a regenerating act from the world above. Thus, one can become part of the organic world by evolving physical life and part of the spiritual world by becoming a Christian. The Christian was a divergent species from the merely human.[42] Entry to each world is not automatic and takes place according to what Darwin called 'descent with modification'. As Drummond put it: 'Some mineral, but not all, becomes vegetable; some vegetable, but not all, becomes animal; some animal, but not all, becomes human; some human, but not all, becomes Divine.'[43]

Since Drummond was a preacher as well as a scientist, a number of analogies from nature were used to convey the point of homilies. The doctrine of atonement – that simply believing in Christ's dying for one's sins is enough to save – was described as a 'molluscan shell'.[44] It was in fact a method of warding off spirituality, by not allowing oneself to be challenged and thus being transformed. Equally, mechanic churchgoing without mental involvement was described as 'parasitism'.[45] The preacher would prepare some spiritual truths for the congregation, but this pre-digested sustenance was not enough to enable spiritual growth. This argument can be linked with the biologist E. R. Lankester's (1847–1929) view that an organism securing food without work injures itself. In general, *Natural Law in the Spiritual World* offered an imaginative scheme for asserting that the laws of science were no threat to Christianity, and in fact tacked religion on to recent developments in science. It did not engage with Darwinism at a deep level, since it simply ignored any intellectual contradictions. But in stating that the natural and the spiritual laws were identical, it accepted that Darwinism could be used as an insight into the mind of God.[46]

Drummond's Christian Darwinism

Even Wallace, the co-discoverer of evolution, had difficulties in coming to terms with the theory applying to humans. He shifted his position on this a number of times, torn between the desire to be evolution's leading light after the death of Darwin in 1882 and his personal views, which, after 1865, included spiritualism.[47] This provided a useful starting point for Drummond's second book *The Ascent of Man* (1894), which directly and explicitly sought to reconcile Darwinism and Christianity. At the beginning, Drummond quoted Wallace, to the effect that it was incontestable that evolution accounted for the development of humans' bodily structure and that we shared an ancestor with anthropoid apes.[48] However, with regard to the full capabilities of humans, Darwinism remained a theory only and could still be modified to match the facts, and future facts coming to light.

Drummond saw evolution as a continuation of natural theology. Nature was God's writing in the universe, and all of it was true. As other natural theologians had pointed out, gradual creation in the way evolution described was an infinitely more subtle and logical method than bringing everything into existence at once, which merely smacked of the magician. There was nothing degrading in humans emerging from the animal kingdom. On the contrary, since humans were more advanced, it was a clear exaltation. Nor was there any contradiction between evolution and creation, as evolution was the only *possible* method of creation. Far from evolution abolishing a creative hand, it demanded it. These points had been made by others, too.[49]

Drummond's main argument, however, stepped on to the territory of Darwin and saw meaning and teleology where the discoverer of evolution had seen only facts. As described above, Darwin had explained the moral sense and altruism as a consequence of humans coming together in groups to strengthen their chances of survival. Gregarious animals were also willing to fight and die for each other, and generally looked to each other's welfare.[50] Thus, by breaking up the extreme selflessness of some individuals into its component parts, Darwin was able to explain it in his own terms. Drummond, on the other hand, declared that altruism was the goal towards which the whole process of evolution was working. The self-interest necessary to survive in the struggle for life was the preliminary stage to altruism, or love, that the scriptures demanded. It could best be seen in the maternal instinct. Evolution had created the mother and she embodied the principle of love, which was God's design. The course of evolution had originally been amoral, but in attaining this higher level it became possible to dream of better humans and societies characterised by greater justice (or, for Drummond, the Kingdom of

Heaven). Echoing Darwin, he also stated that the tribe in which members looked to each other's welfare, loving their neighbour, was more 'fit' than the tribe whose members were indifferent to each other's plight.[51] Societies infused with civilised values allowed their members the highest possible intellectual and physical development.

The Ascent of Man consisted primarily of a popularised retelling of evolution, and one needed to delve at least halfway through it before there was much consideration of religious aspects. A Christian angle emerged only towards the very end, but the preceding theist arguments served the purpose of laying the groundwork for it. Drummond explained that there was no need to 'reconcile' Christianity with evolution, as they were in any case one.[52] Evolution was a method of creation with the object of making more perfect living things. Christianity was exactly the same. According to Drummond's contention, evolution worked through love, as did Christianity. No direct engagement was made with the difficulty that, at some stage, humans received a soul whereas other creatures did not, but this would have marked part of the ascent towards a higher destiny. Humans being of greater ability than animals could be used as evidence for this, though it was unfair to quote Darwin selectively saying that he was not able to account for the mind of humans, nor of how life itself originated.[53] In *The Descent of Man* Darwin had given an explanation for the astounding ability of humans, and it would also have been possible to deduce how the mind originated by analogy from Darwin's lengthy treatment of the development of the eye in *The Origin of Species*.[54]

Drummond ignored the natural theological stumbling block of why God allowed the majority of life to be extinguished at an early stage or what the purpose of the evolutionary cul-de-sac of species that became extinct was. Darwin had noted that the majority of eggs that are laid are eaten by other creatures, and he explained that, in the winter of 1854–5, no fewer than four-fifths of birds on his grounds died due to the severe climate.[55] Christianity had never cared much for animals anyway, but with no Fall in Drummond's teleology it was left unresolved why there were such high rates of infant mortality in Victorian Britain.

Another theory for why there was so much suffering in the world had emerged from evolution – namely social Darwinism – but it drew radically different conclusions from natural selection than what was Drummond's focus. Unsurprisingly, therefore, Drummond rejected the theorising of Benjamin Kidd (1858–1916) about social Darwinism.

Kidd had argued that the 'struggle for life' had intense value, even in today's society, whereas Drummond thought the opposite: competition inside a developed society had the effect of making it unravel, with great losses for the individuals within it. Surprisingly, however, he agreed with Kidd that natural selection did not contain any sanction towards either morality or social progress. This seemed odd, given that, for Drummond, the purpose of evolution was progress towards altruism. But when the environment expanded self-consciously to include the divine, which had in any case always been part of it, as humans gained knowledge of God, that changed.[56] Drummond's conception was that evolution worked towards creating better beings, which of course included moral considerations. There was some scriptural authority for this, as Christ had talked about the Kingdom of Heaven 'being within you', and not as something external. Drummond interpreted this saying collectively, noting how the blade of the corn was visible today, the ear would come tomorrow and the full growth awaited our children's children.[57] While rejecting social Darwinism, Drummond was happy to use Spencer's writings for help in understanding evolution.

These books did not plough deep into Darwinism, nor were they perhaps intended to. Drummond was not just a scholar, but also a preacher. To the extent that his readers were convinced Christians, they were simply looking for a way to hold on to their faith in a scientific age. If someone was able to bring new scientific findings, especially evolution, within the purview of faith, that was a good enough reason to continue as before. It helped readers to avoid the major personal crisis of losing faith, to which Victorians were so prone.[58] In many cases it was not every single tenet of Christianity that the individual felt bound to preserve, but the moral and habitual system of hope in a better life and justice which that religion held out. Consequently, it mattered little if a few of the dogmas of the Church, such as the Fall, had to be abandoned. Even if humans were not inherently tainted by original sin, most people recognised themselves as personally having sinned, and Christ's passion still had meaning as expiation of that. Or the Fall could, paradoxically, be reinterpreted as the stage at which humans become separated from animals in gaining a conscience and thus realising that their selfish actions caused difficulties for others. Did not Genesis talk of the tree of knowledge, whose fruit would allow humans to become like God in knowing good and evil?

Drummond was clear that there was spiritual selection in that all humans would not attain Heaven. Since his belief was that the laws of the spirit world were the same as those existing in the natural world, they could be seen as paralleling the law of natural selection. That natural selection implied that individuals had to look out for themselves, and that this care would be extended by reproductive success to the individual's family, made some sense as the beginnings of altruism. It was important for the conception of the world that Drummond held that social Darwinism could have no import in his system. Social Darwinism used the same or similar facts derived from the theory of evolution to build up a secular and egotistical schema. It was perhaps possible to combine social Darwinism with Calvinism or other denominations supporting double predestination, but while Drummond was a fervent Protestant, his theology was inclusive and meant to appeal to all Christians.

Since Drummond claimed that the authors of evolution and of Christianity were one and the same, he had implicitly accepted that God was responsible for the many failures of individuals in both the animal kingdom and among humans to survive. He never focused on this difficulty levelled against the Christian faith. But was this so difficult to explain away by using the tools at the disposal of the Christian? Suffering was clearly a part of existence, both for those who were successful and survived and for those who led short lives. It seemed perhaps unfair that some should have to suffer purely so that *others* could improve themselves, but was not this principle already inherent in a divided afterlife? By instituting immortality and an afterlife, God had inevitably caused some to suffer in order that others should be rewarded. And given that punishment for the wicked was eternal, this suffering was infinite, whereas Darwin had pointed out that death through nature generally came quickly and involved only limited pain.[59]

Moreover, in a general sense, the historian David d'Avray has shown that any value system is immune to challenges from outside it. That is even the case with a very primitive system such as divination.[60] In comparison, Christianity has hundreds of pages in its Holy Book that could be used to interpret new or challenging facts in such a way that they were brought within the boundaries of the faith. Because Drummond wrote for Christians, he did not require scaling great intellectual heights in order to reassure them and defend their ideology. But although he probably came from one of the traditions most conducive to reconciling science and religion, to be sure, there were less intellectual Christians who 'felt in their bones' that Darwinism was dangerous as a rival to their world view.

In May 1895, Drummond was subject to a motion of censure from the Highlander Calvinists, who disliked the liberalism of the Lowlands and thought *The Ascent of Man* dishonouring to God and degrading to humans. In Drummond, as an evolutionist, they had the perfect target for frustrations that went wider than a single person's unorthodox version of Christianity. The General Assembly voted by 274 to 151 in favour of Drummond after he had been defended in a speech by the principal, Rainy.[61]

Ultimately, for those who *wanted* to hold on to their faith while orienting it towards the most crucial scientific discovery for decades, if not a century, Drummond satisfied. There were clearly many in that category, as attested by the success of his public lectures and books. For people who were sceptical or hostile to either Christianity or Darwinism, his contentions were probably not very convincing. To show this, his findings could be related to the difficulties that evolution posed to Christianity. Many objections remained unanswered, such as how the soul could be implanted in the first human when it was denied to that individual's parents, whose heredity had been mixed to produce him or her. Darwin made clear that evolution happened extremely slowly, so that offspring would always resemble their parents greatly. There simply would never come a point at which it was possible to say that members of a new generation were so advanced, compared with those of the previous one, that they deserved souls.[62] Equally importantly, there was no engagement with the important Christian concepts of the Fall and original sin.

Aubrey Moore's critique of Drummond and his own reconciliation

A second reason why Drummond might be judged to have failed intellectually is that it was possible to do a better job reconciling Darwinism and Christianity. It is at this stage that we would like to bring Aubrey Moore back into the discussion. Moore was one of the critics of Drummond. *The Ascent of Man* did not attain the same level of success as *Natural Law in the Spiritual World* and, probably because it was more specific, came under heavy fire from scientists and theologians alike.[63] Moore rejected as unwise Drummond's reinterpretation of Christianity to fit evolution, because there is always scientific progress and Darwinism may be replaced by a more accurate theory in the future. Moreover, Drummond's version was not the Christianity of the Bible or the received tradition.[64]

Beyond these general objections, Moore criticised Drummond for abandoning the moral sphere by subordinating it to the natural world. Since moral laws have a clear God-given purpose they actually take precedence over natural laws, which are uniform and hence do not tell us that much of theological significance.[65] Similarly, Drummond minimises free will (through concentrating too much on scientific laws), again downplaying the moral laws and thereby also losing sight of the grace of God.[66] The result is the continuation of a religion without its moral core, and such an argument is fatal to the defence of Christianity. It was correct that Christ saw certain parallels between the spiritual world and nature – using the parable of the mustard seed as standing for the Kingdom of Heaven, for instance. But by abandoning revelation, Drummond loses perfect and infallible knowledge of the spiritual world in favour of making theology subordinate to science. These were fair arguments from a Christian perspective, but in order to vindicate them Moore himself would have to overcome the challenges posed to Christianity by Darwinism. Otherwise, he would merely be carping from the sidelines and hardly advancing the debate much.

Like Drummond, Moore accepted that the creation of humans had come about through an evolutionary process. What had become known as 'special creation' had only been launched by Carl Linnaeus in *Philosophia Botanica* (1751) and thus was not essential to Christianity. In fact, in *De Genesi ad Litteram* St Augustine (354–430) had put forward an early theory of evolution.[67] By accepting that God worked according to laws, instead of being autocratic or capricious, we do not thereby take away from his power or majesty.[68] Evolution, though only a theory, was infinitely more Christian than 'special creation' because it implied God's immanence in nature and the omnipresence of his creative power.[69] A contrary theory of occasional intervention conversely meant that he was ordinarily absent. In the words of *The Gospel According to John*, the world was created through the divine Logos, which is why there is unity in nature. Humans, despite Darwinism, were created in the image of God because we share in this Logical nature – a point that Moore attributes to St Athanasius (*c*.296–373).[70] As for nature not being benign, as Darwin had shown, that only put an end to the argument for design, which Immanuel Kant had rejected at an earlier stage anyway. The argument for design had been Darwin's reason for believing in God, but when he substituted natural selection for teleology in the manner of Paley, he stopped believing in God altogether. Therefore, one may add, it might be dangerous to base one's entire religious system upon Darwin's findings, as Drummond was doing.

Moore, however, also left unanswered the problem of suffering, which Darwin had compounded, and the problem of why humans but not animals have souls if they were originally indistinguishable. His answer to both these objections was that a Christian theologian is not bound to know. Since creation was the work of a benevolent God who was wholly good, all his creatures shared in his mercy, according to Moore. God was omniscient, not neglecting even 'a sparrow that falls to the ground', and there was a purpose to and design behind everything.[71] This was merely restating Christian doctrine without engaging in the argument. For Moore, the problem of the soul served only as an argument against traducianism – the doctrine that souls are inherited from parents to offspring – which he noted that orthodox Lutherans believed.[72] As a Catholic, this was of no concern to him. Of all God's creatures, only humans had a relationship with him, and this had been represented by the Bible and the Church as 'likeness'. Humans were self-conscious, free personalities, Moore claimed, and he did not engage with Darwin's argument, outlined above, that since various animals dream and have memories they too may be self-conscious. As mentioned, another difficulty for Christians was that there seemed to be no Fall. Was the Fall in fact a step forward in evolution? Moore thought that the Fall meant a change in humanity's relationship with God and a change for the worse.[73] He had no explanation for it, merely confining himself to stating that Darwinism as a positive science was neither a help nor a hindrance in finding an answer to this conundrum.

Thus Moore, like Drummond, simply ignored problems that were too difficult for him to answer. He was less willing to allow Darwinism to have a role in interpreting Christianity, preferring to keep it merely as a theory of the origins of humans. Moore successfully showed that evolution was compatible with Christianity in the sense that descent from animals should not impinge on faith. It was often God's purpose to use the lowly to accomplish his purposes. The biblical metaphor was that humans had been formed from the dust of the ground.[74] The chosen people were descended from 'a Syrian ready to perish', and were the fewest of the nations around them. They were continually reminded that they had been bondsmen. Yet, the Hebrews were invested with the breath of God and it was they who had been favoured with his oracles. They became the teachers of the world in religious matters. They were perhaps the first humans to have souls, because, in a later essay included in the volume, the soul is defined as the conscious relationship with God. This would, however, mean that not all humans had souls and yet were interfertile with those that did. Darwin had shown that in his day there were many tribes who had no conception of a God. Despite his own thinking leading to illogical results such as this, Moore believed that God was knowable and that science could reveal some of his purposes. Science therefore had some value beyond the merely practical, and it was wrong for believers to try to magnify God by stating that he was unknowable.

Conclusion

We have investigated two attempts to engage with Darwinism by Christian thinkers in late Victorian Britain. Drummond's endeavour may be labelled 'positive' because he sought to formulate Christianity in a different light, using evolution, whereas Moore's engagement was 'negative' in that it merely sought to give Darwinism its limited due and then preserve as much of traditional theology as possible. Both had some reasonable arguments. If there is a spiritual world, logically, it should be connected with the natural world, as both must be the creations of God. The principle of uniformity that is present in nature, meaning that distinct phenomena are in fact connected, makes it not unlikely that the same principles should be present in any spiritual world, too. The leap of faith lies in believing that such an alternate dimension exists at all. Moore was convincing on how Christianity does not need the theory of special creation for it to be believable. He was still able to assert the importance of morality and religion as God-given, separate from how life is modified through natural selection or Lamarckism, which he believed formed part of the popular conception of Darwinism.[75]

Although Darwinian findings drew attention to the problem of suffering and exacerbated it, paradoxically they also invested it with some meaning. This became the doctrine of social Darwinism, which explained that, although there was great suffering and much loss of life, the end result was the continual improvement of species. Human beings have made great leaps forward only because the less well suited have been prevented from contributing greatly to the common gene pool. As noted, Drummond disagreed with the social Darwinism of his time but, logically, he must have been willing to accept that it had had a role in the past. He seems to have embraced this. It was perhaps a dangerous doctrine for a Christian thinker to accept. Had not Christ taught that the meek and the poor in spirit were blessed? Yet Christ's indifference to the world implied that he did not consider material or reproductive success to be anything for which it was worth striving.

While natural selection and sexual selection, considered more comprehensively in Darwin's second book, may indeed have driven us to new heights as a species, from a theological point of view that may be relatively unimportant. It is not certain that it was any easier to be saved in the nineteenth century than it was in the fifth, for all the great improvements that had come about in technology, science, hygiene and material comforts. Darwin thought that morality had become vastly better since ancient times, and had continued to improve during his own lifetime. But if average standards of morality were lower in earlier periods, the example of Christ's demeanour and ethics would have been all the more powerful. In the final analysis, Christian religion has only a limited interaction with induction and deduction because at all stages one has to take account of scripture, with its variegated and sometimes contradictory stories. But belief needs to have some relation to the real world, and it is likely that especially Drummond and also Moore helped preserve faith among Christians who were rational.

Notes

[1] G. Beer, 'Introduction', in C. Darwin, *On the Origin of Species* (Oxford, 2008), p. xxi.
[2] P. J. Bowler, *Monkey Trials and Gorilla Sermons. Evolution and Christianity from Darwin to Intelligent Design* (Cambridge, MA, 2007), pp. 30–1.
[3] R. Fox, 'The University Museum and Oxford Science, 1850–1880', in M.G. Brock and M.C. Curthoys (eds), *The History of the University of Oxford*, vol 6 (Oxford. 1997), pp. 657–8.
[4] Fox, 'The University Museum', p. 642.
[5] Fox, 'The University Museum', pp. 652–3.
[6] R. England, 'Aubrey Moore and the Anglo-Catholic Assimilation of Science in Oxford' (unpublished PhD thesis, University of Toronto, 1997), 61–2.
[7] J. R. Moore, *The Post-Darwinian Controversies. A Study of the Protestant Struggle to Come to Terms with Darwin in Great Britain and America 1870–1900* (Cambridge, 1979), p. 89. In 1896, Temple became Archbishop of Canterbury.
[8] Clergymen who followed Oxford Movement practices were known as 'Ritualists'. See George Herring, *What was the Oxford Movement?* (London, 2002), p. 92.

9. England, 'Aubrey Moore', 86.
10. D. W. Bebbington 'Henry Drummond, Evangelicalism and Science', in T. E. Corts (ed.), *Henry Drummond: A Perpetual Benediction. Essays to Commemorate the Centennial of His Death* (Edinburgh, 1999), p. 23.
11. D. N. Livingstone, *Dealing with Darwin. Place, Politics, and Rhetoric in Religious Engagements with Evolution* (Baltimore ML, 2014), pp. 28–9.
12. Livingstone, *Dealing with Darwin*, p. 29.
13. Livingstone, *Dealing with Darwin*, p. 27.
14. P. Hinchcliff, 'Religious Issues, 1870–1914', in M. G. Brock and M. C. Curthoys (eds), *The History of the University of Oxford*, vol 7 (Oxford, 2000), p. 105.
15. Hinchcliff, 'Religious Issues', p. 106.
16. England, 'Aubrey Moore', 98–100.
17. D. W. Bebbington, 'Henry Drummond, Evangelicalism and Science', in *Scottish Church History Society Records*, vol. 28 (1998), p. 140.
18. J. R. Moore, 'Evangelicals and Evolution: Henry Drummond, Herbert Spencer and the Naturalisation of the Spirit World', *Scottish Journal of Theology*, 38/3 (1985), 391.
19. C. Lennox, *Henry Drummond. A Biographical Sketch* (London, 1905), p. 94.
20. H. C. G. Matthew and B. Harrison (eds), *Oxford Dictionary of National Biography*, vol. 16 (Oxford, 2004), p. 958.
21. Bebbington, 'Henry Drummond', in Corts (ed.), *Henry Drummond*, p. 38.
22. D. Alexander, 'Creation and Evolution', in J. B. Stump and A. G. Padgett (eds), *The Blackwell Companion to Science and Christianity* (Chichester, 2012), p. 234.
23. A. E. McGrath, *Darwinism and the Divine* (Chichester, 2011), p. 171.
24. C. Darwin, *The Descent of Man and Selection in Relation to Sex* (London, 1909), p. 40.
25. Darwin, *Descent of Man*, p. 53.
26. Darwin, *Descent of Man*, p. 100.
27. Darwin, *Descent of Man*, p. 107.
28. Darwin, *Descent of Man*, p. 113.
29. Darwin, *Descent of Man*, p. 127.
30. Darwin, *Descent of Man*, p. 936.
31. Darwin, *Descent of Man*, p. 180. In the context of the American Civil War, Darwin rashly wrote that abolition might be worth 'a million horrid deaths' in the long run. See A. Desmond and J. Moore, *Darwin's Sacred Cause. Race, Slavery and the Quest for Human Origins* (London, 2009), p. 326.
32. Lennox, *Henry Drummond*, p. 40.
33. Moore, 'Evangelicals and Evolution', 386.
34. Lennox, *Henry Drummond*, p. 88.
35. Lennox, *Henry Drummond*, p. 82.
36. Bebbington, 'Henry Drummond', in Corts (ed.), *Henry Drummond*, p. 30.
37. A. C. Cheyne, 'The Religious World of Henry Drummond (1851–1897)', in Corts (ed.), *Henry Drummond*, p. 9.
38. Wells argued that we can move along the three physical dimensions: length, breadth and thickness. The fourth dimension is time. In just the same way as an air balloon will allow someone to travel 'up' against gravity, so eventually a machine can be built that allows travel against the general direction of time: H. G. Wells, *The Time Machine: An Invention* (Jefferson NC, 1996), chapter 1.
39. H. Drummond, *Natural Law in the Spiritual World* (London, 1907), p. xiii.
40. Drummond, *Natural Law*, p. 214. A later student of Drummond's work, Joseph Needham,

also regarded this analogy as far fetched: J. Needham, 'The Naturalness of the Spiritual World. A Reassessment of Henry Drummond (1939)', in J. Needham, *Time. The Refreshing River* (Nottingham, 1986), p. 29.

41 Drummond, *Natural Law*, p. 404.
42 A. Scott, '"Visible Incarnations of the Unseen". Henry Drummond and the Practice of Typological Exegesis', *British Journal of the History of Science*, 37/4 (2004), 442.
43 Drummond, *Natural Law*, p. 412.
44 Drummond, *Natural Law*, p. 431.
45 Drummond, *Natural Law*, p. 350.
46 An admirer of Drummond gave as an example John 12:24: 'Except a grain of wheat fall into the earth and die, it abideth by itself alone, but if it die it beareth much fruit': in T. Hunter Boyd, *Henry Drummond. Some Recollections* (London, 1907), pp. 65–6.
47 D. Stack, *The First Darwinian Left. Socialism and Darwinism 1859-1914* (Cheltenham, 2003), p. 25.
48 H. Drummond, *The Lowell Lectures on the Ascent of Man* (London, 1894), p. 8.
49 For instance, by Robert Chambers in *Vestiges of the Natural History of Creation* (1844).
50 Darwin, *Descent of Man*, p. 154.
51 Drummond, *The Lowell Lectures*, p. 306.
52 Drummond, *The Lowell Lectures*, p. 438.
53 Drummond, *The Lowell Lectures*, p. 156.
54 C. Darwin, *On the Origin of Species by Means of Natural Selection or the Preservation of Favoured Races in the Struggle for Life. Introduced by Richard Keynes* (London, 2006), pp. 148–9.
55 Darwin, *Origin of Species*, p. 53.
56 Drummond, *The Lowell Lectures*, pp. 68–9.
57 Drummond, *The Lowell Lectures*, p. 444.
58 The most marked being perhaps that of Sir Charles Lyell, whose *Principles of Geology* (1830) had shown how the world was millions, not thousands, of years old: Moore, *Post Darwinian Controversies*, p. 105.
59 Darwin, *Origin of Species*, p. 61.
60 D. L. d'Avray, *Rationalities in History. A Weberian Essay in Comparison* (Cambridge, 2010), pp. 83 and 88–9.
61 Livingstone, *Dealing with Darwin*, p. 22; Lennox, *Henry Drummond*, p. 93. In May 1892, Drummond, along with three other professors, had also been accused of heresy. Drummond repudiated the accuracy of the reports, and all four men were cleared of the impeachment charge at the General Assembly. See Lennox, *Henry Drummond*, p. 86.
62 This is an interesting application of Zeno's paradox of Achilles and the tortoise. Achilles never catches up with the tortoise because it has a head start. As soon as Achilles has covered that distance, the tortoise has moved a little further. This continues infinitely. While Achilles is faster than the tortoise and should therefore outdistance it, he does not. While human beings deserve a soul due to our advanced abilities, no one generation was so much more advanced than the previous generation that a soul could be implanted.
63 Bebbington, 'Henry Drummond', in Corts (ed.), *Henry Drummond*, p. 38.
64 A. L. Moore, *Science and the Faith. Essays on Apologetic Subjects. With an Introduction* (London, 1889), p. 2.
65 Moore, *Science and the Faith*, p. 8.
66 Moore, *Science and the Faith*, p. 28.
67 Moore, *Science and the Faith*, pp. 175–6.

68 Moore, *Science and the Faith*, p. 181.
69 Moore, *Science and the Faith*, p. 184.
70 Moore, *Science and the Faith*, p. 185.
71 Moore, *Science and the Faith*, p. 199.
72 Moore, *Science and the Faith*, p. 210.
73 Moore, *Science and the Faith*, p. 221.
74 Moore, *Science and the Faith*, p. 205.
75 Moore, *Science and the Faith*, p. 163.

FROM BIOPOLITICS TO EUGENICS: THE ENCYCLICAL *CASTI CONNUBII*

Emmanuel Betta

Historical research has often put forward the view that Catholics in general, and the Roman Catholic Church, in particular, were the only organized opposition to eugenics, both on a legislative and practical level. The turning point in the Catholics' attitude towards eugenics has been identified as occurring with the publication of the encyclical *Casti connubii*, issued on 31 December, 1930 by Pope Pius XI (1857–1939). This document denounced birth control, abortion and sterilization, contested as ways to interfere in the order of nature through the control of reproductive sexuality. It is this encyclical, especially its doctrinal statement on sterilization, which is considered by many eugenics historians to have driven Catholics from different national backgrounds to adopt a more sharply critical attitude towards eugenics practices and laws.[1] This change was not immediate and it varied widely from country to country. However, a large number of Catholics seemed to reject sterilization and to firmly oppose to different forms of eugenics. Moreover, the doctrinal realignment of Catholicism to the denunciation outlined in the *Casti connubii* was not uniform with regards to sterilization either. In Germany, two intellectuals, Hermann Muckermann (1877–1962), a former Jesuit and later professor and director of the Eugenics section of the Kaiser Wilhelm Institute for Anthropology, Human Heredity and Eugenics (Das Kaiser-Wilhelm-Institut für Anthropologie, menschliche Erblehre und Eugenik) in Berlin, and the priest Josef Mayer (1886–1967) continued to endorse eugenics and, in Mayer's case, even sterilization and euthanasia.[2] Due to widespread favourable opinions on eugenics and sterilization among American Catholics, the impact of *Casti connubii* was stronger in the US. After the encyclical's publication, two Catholic priests, John A. Ryan (1869–1945) and John Montgomery Cooper (1881–1949), resigned from their important positions in the American Eugenics Society, while Catholic clergy put forward a stronger opposition towards eugenics laws all over the country.[3] On the other hand, historians have often stressed the ambiguity of the Catholic clergy towards eugenics particularly because Catholicisim and eugenics have so many elements in common. In particular, it should be noted that even the condemnation of sterilization in *Casti connubii* was not unconditional and it can be mainly explained as the result of a wider overview on reproduction and the social forms of its expression and control, which is part of the broader Catholic doctrine relating to the body.

Casti connubii made quite a strong impact on Catholic discourse concerning reproduction and the, discipline of sexuality and eugenics, but it did not put a stop to discussions on wider issues of eugenics'. On the one hand, it should be stressed that this encyclical did not condemn eugenics *per se*, but condemned specific practices that in different ways could be described as *negative eugenics*, as already noted.[4] On the other hand, *Casti connubii* provided a clear and strict disciplinary structure for dealing with issues concerning reproduction and sexuality, by means of which bio-political questions which had been discussed in Catholic circles since the mid-19th century could be interpreted systematically'. Thus, the eugenics encyclical was seen as one aspect within a growing bio-political attitude put forward by secularized institutions, mainly the state, and their interventions for governing family and marriage, through the control of the human body and reproduction. From this point of view, the Catholic condemnation of eugenics appears to be more deeply directed towards the state than towards eugenics itself.

The term eugenic, or a derivative of the term, appears five time in the text of the *Casti connubii* and on each occasion it is used together with other adjectives to indicate one of the many ways of legitimizing therapies concerning reproduction, such as contraception, abortion and sterilization. All of these cases dealt with interventionist therapies which had long been the object of a normative and doctrinal intervention of the Catholic Church, mainly through the action of the Congregations, in particular the Congregation of the Holy Office, otherwise known as the Roman Inquisition.[5] In many ways the second half of the nineteenth century represents a decisive turning point in the history of Catholic moral discourse. For its part, the Catholic Church began to react to civil marriage, producing documents that set out the sacramental character of marriage, seen as a prerequisite to the claim of exclusive management of marriage put forward by the Catholic Church itself. The Congregation of the Holy Office similarly started to outline in detail practices and behaviours relating to marriage, sexuality and reproduction. While the path of the former was set on a progressive restriction, given the irreversibility of the process of legalization of family and marriage by the state, disciplining biomedical practices followed a different path. In its role of restoring the primacy of the Catholic Church and religion within society and culture, the Holy Office considered matters relating to nature, the body and health, as strategic. By disciplining medical practices, the centrality of Catholic morality and discipline within the public arena could be reaffirmed. In this way, the disciplining activity of the Holy Office expanded significantly.[6]

Between 1816, when *Penitenzieria apostolica*, the Congregation responsible for issues relating to' the discipline of auricular confession, first discussed contraception, and 1901, when the Holy Office produced the last decree on therapeutic interruption of pregnancy, Roman Congregations discussed many issues. These included magnetism and hypnosis from the 1840 onwards; cremation between 1886 and 1892; birth control and contraception after 1851; artificial insemination after 1877; sterilization after 1885;[7] obstetrics and therapeutic interruption of pregnancy after 1884 and Darwinism from 1877 onwards.[8] Then, at the beginning of the twentieth century, the members of the Holy Office began to direct their attention to eugenic practices. *Casti connubii* can be considered at the same time as the final outcome of a long historical process of formulation of the Catholic doctrine concerning medicine, reproduction and sexuality deeply rooted in the cultural and political context of the nineteenth century, and as a product of the earlier decades of the twentieth century. The combination of these two paths produced a very important document for Catholic discourse on sexuality and reproduction, making it highly influential in many ways right up to the present day. It has established the prevalence of a rigorist doctrinal interpretation that, since then, has deeply influenced the Catholic approach to these issues. Contrary to what was expected, in 1968, *Humanae vitae*, the second important bio-political encyclical of the Catholic Church issued by Pope Paul VI (1897–1978), condemned the use of the pill, and this was done precisely in order to avoid contradicting the statements of *Casti connubii*.[9]

This article argues that the multifaceted conception of reproduction and eugenics put forward in *Casti connubii* can be fully understood if one considers two kinds of processes throughout the nineteenth century that prepared the cultural and political background to which the encyclical made explicit reference. Firstly there is the deep and intense action of disciplining medical practices linked to reproduction which had been practised by the Congregation of the Holy Office since the middle of the nineteenth century; secondly here are the cultural and political reactions of the Catholic hierarchies to two traumatic historical events: the publication of Charles Darwin's *On the Origin of the Species* in 1859 and the end of the Pontifical State, together with the loss of political power following the foundation of the Italian Kingdom in 1861. Both events were perceived as affecting and reducing the role played by the Catholic Church in society and culture, mainly because political sovereignty seemed increasingly to assert its power through the definition and the control of natural and biological aspects of humans. What was at stake here, then, was controlling the truth and meaning of life, and consequently, fighting against the state's increasing authority over reproduction and sexuality'.

The Catholic Church reorganised its hierarchy on both a political and cultural level. From a political point of view, in 1870 the first Vatican Council approved the dogma of papal infallibility when the Pope speaks *ex cathedra*, that is to say when he speaks as pastor and doctor of all Christians and defines a doctrine concerning faith and customs. As a result of this decision the Pope's words bind all members of the Catholic Church, at least in the public sphere. On a cultural level Pope Leo XIII (1810–1903) issued the encyclical *Aeterni patris* in 1878, which in fact reinstated the theology of Thomas Aquinas (1225–74) as the theology of Catholicism. Neo-Thomism was conceived to give strength and unity to the Catholic discourse in all aspects of life and culture. This approach had an important impact on medical and natural science, because the revival of Thomas Aquinas's theology also entailed a revival of Aristotelianism and Hylemorphism as the most appropriate methods of understanding the order of nature. This implied adopting the relationship between form and matter as the main cognitive scheme of investigation and also a hierarchical conception of knowledge, where medicine was reduced to the state of an art unable to produce principles and values in the way they were defined by theology. This political and cultural atmosphere influenced the manner in which the relationship between body, cultures and reproduction has been conceived both by the norms produced by the Roman congregations and subsequently by the encyclical issued by Pope Pius XI.

From the mid-nineteenth-century onwards, Catholic hierarchies have progressively perceived biomedical intervention relating to the body of the individual and of the population as a crucial danger. Biomedicine started to be considered as one of the main vehicles for secularization, and Catholic hierarchies increasingly focused their moral and doctrinal discipline on defending the natural integrity of life and the body both seen as threatened by evolutionary science and biomedicine. For the Catholic Church, the political importance of the battle for reproductive control had already emerged in the encyclical *Rerum Novarum*, issued by Pope Leo XIII in 1891 and dedicated to the social question. It founded the so-called social doctrine of the Catholic Church, which promoted the necessity of an agreement between capital and labour based on Christian solidarity. In the first preparatory versions of this encyclical, written by the Jesuit theologian Matteo Liberatore (1810–92), a long section on birth control practices – referred to as 'onanism' – was drafted. The aim was to fight the political suggestion of restrictive laws concerning marriage or the practice of making a marriage sterile, deemed to be 'neo-Malthusian', as the application of Thomas Robert Malthus's (1766–1834) economic theories to society and population came to be known. Standing out against this political trend, Liberatore proposed the definition of marriage as a primordial and natural right and emphasizing the duties of the individual in the domestic society. Although the section on onanism did not appear in the final version of *Rerum Novarum*, it is important to stress that it was a similar approach that would characterise the argument put forward forty years later in *Casti connubii*.[10]

Casti connubii was dedicated to Christian marriage, and the main reason for its advocacy is usually identified as the 1930s decision of the Anglican Bishop Conference in Lambeth to partially endorse the use of contraception.[11] This issue had been discussed throughout the nineteenth century by the Roman congregations. Initially the *Penitenzieria apostolica* adopted a tolerant approach, accepting implicitly the use of contraceptive methods by Catholic couples with many children wanting to control their own fertility for economic reasons, that is to say the inability to sustain more children. Then, after 1851, the Holy Office intervened many times, imposing a rigorist approach to Catholic confessors, insisting on a commitment for a more detailed interrogation in relation to reproductive behaviours and on the duty of a clear condemnation of contraceptive practices.[12]

Casti connubii promoted a clear and rigorist interpretation of marriage based on the teachings of St Augustin, the Council of Trento and Pope Leo XIII. Infused with Augustinian theology, the encyclical takes for granted the three blessings of marriage – *proles, fides, sacramentum* (offspring, conjugal faith and the sacrament) – as the obligatory references for the definition of relationships and hierarchies within marriage as well as for stating the dominant position/priority given to the natural order of procreation and the subordination of a woman to her husband. This pattern was based on a negative interpretation of sexuality, also derived from Augustinian anthropology, which considered sexuality as a hazardous force to be controlled within the order of morality and procreation. This argument was accompanied by the usual alarmist references to a cultural and political context dominated by sexual freedom and public control of reproduction. The encyclical promoted the model of a pre-industrial Christian patriarchal family with a high birth rate and in which the woman is subordinated to her husband and entirely focused on family life and child-care. Together with the denial of any possibility for the emancipation of women – concerning work, personal property or administration – *Casti connubii* recommended increasing salaries, to allow the father, as the only breadwinner, to sustain his own family.

The indissolubility of Christian marriage was presented as the cornerstone of the Christian family. At the same time, the Christian concept of family, with its internal hierarchies, was shown to underpin an organicist concept of social relations, inside and outside the family, and society. The family, 'as more sacred than the State', was considered non-alterable, because its structure and its fundamental laws were 'established and confirmed by God', and they 'must always and everywhere be maintained intact'.[13] This perspective anchored the family and its internal relationship to the level of natural order and laws and meant that it was not, as such, modifiable by any choice, be that of the individual or of the state. Conceived in this way, this rigid moral discipline of marriage excluded any possible intervention into an arrangement considered both natural and moral. One can thus see that the state was criticised for being an institution which did not respect the 'natural' order of marriage and family, as it promoted divorce and civil marriage. This argument was reinforced by citing explicitly the encyclical *Arcanum divinae sapientiae*, issued by Pope Leo XIII in 1880 which focused on the family, with the aim of reaffirming the exclusive legislative and judiciary power of the Catholic Church regarding marriage as opposed to divorce and civil marriage laws. In the same way, *Casti connubii* reiterates the sacramental value of marriage and its subsequent indissolubility in contrast to the modern view of marriage as a contract promoted by the French civil code: 'it is hardly necessary to point out what an amount of good is involved in the absolute indissolubility of wedlock and what a train of evils follows upon divorce'. Therefore, the encyclical referred to the Fascist state as a positive example of relationship with natural order. The Lateran Pact, signed between the Catholic Church and Mussolini in February 1929 is described as:

> a striking example to all of how, even in this our own day (in which, sad to say, the absolute separation of the civil power from the Church, and indeed from every religion, is so often taught), the one supreme authority can be united and associated with the other without detriment to the rights and supreme power of either thus protecting Christian parents from pernicious evils and menacing ruin.[14]

The condemnation of reproduction was a key point of the encyclical's basic argument about reproduction and sexuality from a twofold perspective. The disciplinary argument aimed to contrast both the individual dimension of self-determination of sexual and reproductive life and the institutional action of the state directed at population control and demographic policies. This contrast between the Catholic discourse and the concrete practice of bio-politics expresses a sort of competition between two different paths for defining the semantics of life and for defining the methods and objective of its concrete government. Birth control was one of the main battlegrounds. Using a brief and quite hermetic phrase, *Casti connubii* allowed for the use of the so-called 'natural methods' to control conception, that is to say it allowed for sexual intercourse during the period of female infertility. It was the transposition of an older argument suggested to the Congregation of the Holy Office in 1873 by the Belgian abbot Auguste Joseph Lecomte (1824–81). He followed the theories relating to the regularity of the female period, suggesting the division of the female cycle into a fertile period and an infertile one, laid out in 1845 by the French physician Félix Archiméde Pouchet (1800–72) in his *Théorie positive de l'évolution spontanée et de la fécondation des mammifères et de l'espèce humaine*. Lecomte wrote and distributed a book in northern France and Belgium in which he suggested that confessors should mention this method of contraception as a fully moral solution for Catholic couples to control their own fertility. He then asked the Holy Office for an authoritative approval of his theories and, subsequently, wrote to the inquisitors explaining that a direct confrontation with the news produced by science and medicine was necessary for the pastoral activity of the clergy. Lecomte emphasised the fact that Catholic couples looking for ways to control their own fertility already had the possibility to contact doctors to ask for advice. According to Lecomte, the Catholic Church had to compete with doctors in producing and distributing different advice and solutions for parishioners. The Roman Inquisition examined both Lecomte's request and his book. They found nothing wrong with the arguments and theories he used; it was simply a question of timing. They believed that the diffusion of this kind of solution, in a book written by a priest, could have triggered immoral and degenerate behaviours in the population. So they forced the book to be the withdrawn.[15] More than fifty years later the method proposed by Lecomte was recognized by *Casti connubii*, although in a rather veiled way.

The general argument against human intervention into the natural order of reproduction associated abortion with contraception. This condemnation was reinforced by a quotation from St Augustin about the 'cruel lust of couples avoiding procreation, who consequently could no longer be considered man and wife'.[16] 'The use of the term eugenics first emerged in this part of the encyclical in regards to abortion, and then to sterilization. In both cases the term was used in a direct connection with the state and its intervention in reproduction. Firstly, it indicated various reasons accepted by the state for legitimizing abortion – 'medical, social, or eugenic indication' – and the request for those indications to be validated by public law and then in no way penalized. Talking about those indications, the encyclical presented a complex judgment, implicitly accepting positive eugenics, and rejecting negative eugenic methods:

> What is asserted in favour of the social and eugenic 'indication' may and must be accepted, provided lawful and upright methods are employed within the proper limits; but to wish to put forward reasons based upon them for the killing of the innocent is unthinkable and contrary to the divine precept promulgated in the words of the Apostle: Evil is not to be done that good may come of it.[17]

Casti connubii employed the term eugenic in relation to sterilization on three other occasions. There were several points to this argument. On the one hand, the verdict was strongly negative based on the argument of the inviolability of the natural order including the human body, as created by God. On the other hand the encyclical mentioned the question of the offspring's wellbeing, assuming that this was a morally acceptable objective:

> there are some who over solicitous for the cause of eugenics, not only give salutary counsel for more certainly procuring the strength and health of the future child – which, indeed, is not contrary to right reason – but put eugenics before aims of a higher order, and by public authority wish to prevent from marrying all those whom, even though naturally fit for marriage, they consider, according to the norms and conjectures of their investigations, would, through hereditary transmission, bring forth defective offspring. And more, they wish to legislate to deprive these of that natural faculty by medical action despite their unwillingness; and this they do not propose as an infliction of grave punishment under the authority of the state for a crime committed, not to prevent future crimes by guilty persons, but against every right and good they wish the civil authority to arrogate to itself a power over a faculty which it never had and can never legitimately possess.[18]

Also in this case, there was a clear reference to the role of the state and its institutions, as 'where no crime has taken place and there is no cause present for grave punishment, they can never directly harm, or tamper with the integrity of the body, either for the reasons of eugenics or for any other reason'.[19]

The general argument concerning abortion and sterilization, which referred to medical indication for intervention into the natural order of reproduction, has direct connections with the norms concerning medical activities and therapeutics produced by the Congregation of the Holy Office during the second half of the nineteenth century. With regard to the condemnation of abortion, *Casti connubii* explicitly adopted the decrees produced by the Congregation of the Holy Office in 1884, 1895 and in 1898. The 1884 decree was the first one to deal with the therapeutic interruption of pregnancy; it was reached after two years of discussion between inquisitors who initially discussed the opportunity of an explicit answer, then content of the condemnation of the practice. In the end, they established that it was not correct to teach that embryotomy (a form of pregnancy interruption) could be employed to treat a pregnant woman whose life was in danger. The 1895 decree established that procured abortion could not be employed, even to avoid the pregnant woman's death. The decree of 1898 extended this prohibition to include cases of extra-uterine pregnancy. The interrogations arrived to the Holy Office mainly from medical catholic universities and theological seminars in France and Quebec. They presented a sort of dilemma for Catholics acting in healthcare and obstetrics in reconciling the necessities of therapeutics with the principles of moral theology and catholic doctrine. In 1886, for example, clinicians from the Faculty of medicine at the Catholic University of Lille sent a long and detailed list of sixteen practical obstetrical cases, taken from the practical experiences of the clinicians themselves, requesting to the Roman Inquisitors to apply the 1884 decree, which forbids the teaching of embryotomy in Catholic schools, to those cases. The list included the caesarean section and artificial premature birth, craniotomy, embryotomy, procured abortion, all varying in relation to the age of the unborn child and its viability. The theological discussion revolved around the theological argument of the double effect of a single action, that is the permissibility of an action causing serious harm as a secondary effect of the promotion of a positive goal. Or, in other words, the possibility of applying moral judgment on the first goal pursued – the woman's health – and not on the death of the unborn child which was considered to be as the second instrumental goal of the action. The decree of 19 August, 1889 condemned any surgery killing the unborn or the woman.

This decree, in conjunction with the 1884, 1895 and 1898 decisions, was part of a process of normative production, which started in 1884 and was concluded in 1901, through which the Roman Inquisition set out its condemnation of every therapy concerning pregnancy which was unable to guarantee the safety of the foetus: regardless of the time of the pregnancy or the viability of the foetus. This inquisitorial disciplinary intervention was very important for both defining the condemnation of abortion and clarifying the interpretative assumption for the judgment on eugenics. With regard to abortion, indeed, the Holy Office had to discuss many cases submitted by theologians' seminars and catholic medical faculties mainly from France, Belgium and Quebec, presenting different situations in which physicians considered interruption of pregnancy as the only therapy for saving the life of the endangered pregnant woman. Therefore, this discussion represented the first important occasion in which Roman inquisitors had to face the clash between morality and therapeutics, rather than the relationship between medical and therapeutic indications and concrete modification of the natural order of reproduction. This process produced an implicit subjectification of the foetus itself, appearing in the inquisitorial rhetoric as a subject made fully equal to the woman. The relevant disciplinary statement resulting from these five inquisitorial decrees was the definition of the pre-eminence of the natural order of reproduction as non-alterable by any reason or indication even medical such as the safety of the pregnant woman's life.[20] The statement of the inviolability of natural order of reproduction then became the main argument supporting also the intervention of the Roman Inquisition on artificial insemination and sterilization, and, later on, eugenics.[21]

In *Casti connubii* it was not the first time that Roman Vatican Authorities had discussed about sterilization. The Congregation of the Holy Office had received many petitions from different countries since 1885 – Italy, Quebec, Germany, Brazil – presenting medical cases relating to women, on whom surgical procedures had been carried out, including ovariotomy, oophorectomy or the so called Porro's caesarean section by excision of the uterus and adnexae, a procedure which had been improved by the Italian obstetrician Edoardo Porro (1842–1902) in 1876. It was a fundamental operation in the history of obstetrical surgery, because it drastically reduced the female mortality rates from caesarean section. Nevertheless, it was not immediately accepted because of its practical consequences for the woman's fertility. In making the woman infertile through surgical intervention, this kind of operation was considered problematic from a theological point of view, because it prevented the achievement of the main objective of marriage, that is, procreation. Many Catholic physicians refused to adopt it and argued against Porro. He himself consulted the bishop of Pavia Lucido Maria Parrocchi (1833–1903) for advice about the morality of such an operation. Parrocchi fully approved it, basing his judgment on the idea that the operation did not prevent the possibility of sexual intercourse so it was morally acceptable.[22]

The main argument used within the Roman Inquisition to judge these cases was the concept of impotence in relation to marriage. The inquisitors focused on the kind of impotence, discussing whether this operation was preventing the capacity of sexual intercourse (*impotentia coeundi*) or the capacity to procreate (*impotentia generandi*). The first two interrogations were focused on the therapeutic character of these operations to cure cancer or similar pathologies of the women's reproductive system. In the first interrogation, the bishop of Köln, the Jesuit Paul Ludolf Melchers (1813-95), asked whether the case of a woman whose uterus and ovaries had been surgically removed could be evaluated through the brief *Cum frequenter* issued by Pope Sixtus V (1521-90) in 1587, which declared invalid the marriage of eunuchs and persons lacking the use of both testicles. In the second interrogation, a priest from Faenza, between Bologna and Ravenna, presented the case of a woman, whose uterus and ovaries had been surgically removed, , who was already married, asking if this condition, perfectly known and accepted both by the married couple and the local community, should be considered as an obligation to nullify the marriage. The priest first asked the *Penitenzieria Apostolica*, the Roman congregation in charge of auricular confession, about the validity of the marriage, and then to the Holy Office. The debate among inquisitors was long and complex and they sought the expertise of physicians and surgeons, about the specific characters of surgeries and its consequences for the fertility and sexuality of the woman who had undergone surgery. Edoardo Porro himself was involved in this debate. In his report, he clearly explained that in similar cases a woman who had undergone surgery would definitively become sterile. The Roman inquisitors adopted a moderate approach with regards to the interrogations from Köln (Cologne) and Faenza. They asked the German Bishop to delay the marriage, until a final decision had been taken and they advised the priest in Faenza to not to disturb the couple, because they were ignorant about the validity of the marriage after the woman had undergone surgery.

From the mid-1890s onwards inquisitors were also being consulted concerning the use of sterilization as a method of preventing diseases in both the woman and the offspring. In 1895, the apostolic nuncio Girolamo Gotti (1834-1916), archbishop of Petra, sent an enquiry from Brazil concerning Abele Parente (1851-1932),[23] an Italian doctor who advertised in journals and newspapers his method of contraception, which involved sterilizing women. The origin of this was the condemnation of a book published in Rio de Janeiro in 1893, by Francisco de Castro (1857-1901), who was Professor at the Rio de Janeiro Faculty of Medicine and director of the *Directoria sanitaria de capital federal* and who promoted Parente's proposal.[24] He promoted this method for women affected by diseases that were dangerous to the unborn child; or women who were suffering serious effects as a result of syphilis, insanity, epilepsy, tuberculosis, or cancer; or women married to elderly men who were neuropathic, alcoholic, syphilitic or who didn't want 'degenerated' offspring. Parente's proposal caused a wide public debate, and a polarization between people in favour or against his method. The case was more problematic for the Catholic Church because, according to the nuncio, Parente mentioned an unspecified 'Vatican ecumenical council' supporting his method. The nuncio feared dangers for the spread of this method among Catholics and priests. The inquisitors discussed the De Castro book, condemning it and asking 'its author to withdraw it, but he refused.

Those different norms concerning therapies or medical interventions for reproduction, together with the statement of marriage as a sacrament, were the main cultural premises to the elaboration of *Casti connubii*. Between the end of the nineteenth Century and the end of the papacy of Pope Pius XI marriage became the main focus for Catholic discourse on the family. This was the response to the process of secularization expressed in the spread of marriage when conceived as civil contract, controlled by the State which was considered by the Catholic Church as a sort of pathology of modern times. Together with civil marriage, sexual education was considered by the Catholic Church a strategic area for distinguishing the secular and rebuilding Christian society. This attention was strictly connected with eugenics and the perspective that would have found then the definition in *Casti connubii*. In fact, the education of young people, in particular sexual education, was a major issue which would feature in endeavours to restore Catholicism. The fact that an important Roman Catholic Congregation as the Roman Inquisition decided to discuss sexual education was the sign of the importance placed to this question by Catholic Church.

The first time that Roman inquisitors were openly faced with eugenics had its roots in an interrogation concerning sexual education. The same debate conducing to the first decree on eugenics in 1931 was deeply related with the issue of sexual education in schools for children and in theological seminars for priests. The perspective motivating the radical refusal of every form of sexual education, as well as the refusal of every intervention on the nature of human sexuality was the same. The Church considered sexual education and every medical intervention into sexuality or reproduction as a means by which the state could change the family from within. When seen in this way it was possible to talk about sexuality only in relation to religion, where sexuality had meaning and goals connected to reproduction.

This approach towards the entire field of reproduction and eugenics emerged clearly in the first discussion on eugenics within the Holy Office. In 1925, the Roman inquisition received a book on eugenics issued by the Naples branch of *Azione Cattolica*, the most important Italian catholic association, founded in 1867. The book was the result of a debate organized the previous year to discuss eugenics and to define the Catholic opinion on it. The authors were the Jesuit Giuseppe de Giovanni (1876–1967), Rector of a college in Naples and dean of the Faculty of Theology S. Luigi in Posillipo, and Mario Mazzeo (1889–1973), assistant professor at the Institute of Hygiene of the Royal University of Naples. The foreword was written by Giuseppe Moscati (1880–1927), professor of physiological chemistry and clinical medicine at the University of Naples. He was later sanctified by the Catholic Church. In the preface Moscati recognized that eugenics had 'the high concept of protecting the human race from decadence' but he contested its methods because it is, in part, 'offensive to human freedom, ethics and economics, life or anti-physiological'.[25] The first critical target was the pre-marital certificate. Moscati said that self-restraint was the only remedy for the inheriting disease. In particular, it had to be applied to all the 'unfit' and 'all young people' in order to avoid the 'maximum communicable disease, which is the symbol of original sin, syphilis'. Therefore, according to Moscati, marriage had to be particularly encouraged at an early age as 'physical and moral prophylaxis'.

The authors, de Giovanni and Mazzeo, adopted a similar interpretation of eugenics. They recognized the need for action on the body of society, cleanse it of hereditary defects and degeneration, but they disputed the methodology. Eugenics and the Church's moral discourse had goals in common: the fight against alcoholism, immorality and tuberculosis, the improvement in economic conditions of the worker, the dissemination of the precepts of hygiene. Birth control was the deep difference separating them. According to Mazzeo and de Giovanni, the only acceptable form of eugenics was that of the Catholic Church, which alone had the necessary framework for the purpose of human betterment. The list of remedies which formed the nucleus of a strong and valuable dimension of Catholic eugenics was long and included: pre-marital chastity, the discipline of marriage and sexuality, abstinence as a method of birth control, marriage as prophylactic anti-syphilis measure, breastfeeding, and limitation of marriage between blood relatives. The author's final statement was clear: the only way to connect morality with eugenics was to integrate it within the moral discourse of the Catholic Church.

The Holy Office examined Mazzeo and de Giovanni's book, and entrusted Henri Le Floch (1862–1950) with the task of writing a report. Le Floch was an advisor to the Roman inquisition (*consultore*), a member of the Congregation of the Holy Spirit, director of the French seminar in Rome and an antimodernist having a traditional political position close to *Action française* of Charles Maurras. Le Floch studied eugenic theories in general and Mazzeo and de Giovanni's book together with the book written by the Jesuit Georges Hoornaert (1876–1950) entitled 'Le combat de la pureté. A ceux qui ont vingt ans' (The Fight for Purity. To Those who are Twenty Years Old), and published in 1923 originally in Brussels and Paris by Action Catholique. This book also included a foreword by another Jesuit, the moral theologian Arthur Vermeersch (1858–1936) and was immediately translated into Italian. Vermeersch was deeply involved in the cultural battle against birth control in Belgium and, subsequently, in Rome where he became a professor of moral theology at the Gregorian University and a close collaborator of Pope Pius XI.[26] The definition of *educatio puritatis* (education of purity) was a term adopted by the French pastoral press at the end of nineteenth century to indicate sexual education based on catholic morality. The first author who wrote on this issue was the French abbey Joseph Fonssagrives (1860–1920), chaplain of the *Cercles Catholique des étudiants de Paris* (Paris Catholic circles of students of Paris) and professor at the *Pétite Seminaire* of Paris who published his volume *Conseils aux parents et aux maîtres sur l'éducation de la pureté*, in Paris in 1902.[27]

The Roman inquisition did not condemn the idea of eugenics proposed by the book. In a brief note the consultants wrote: 'nothing was found against faith and customs', and eugenics, 'when is contained within borders and treated with caution, is to be praised'. The decision to discuss this matter publicly was considered detrimental: 'it's really strange that a priest deals with topics concerning the eugenic form for the improvement of the human race!' A second and deeper critic contested the explicit language adopted by Mazzeo and de Giovanni, beginning with the fact that moral theology has always adopted Latin as the language used for discussing issues concerning sexuality. In his report, Le Floch stressed negative implications of eugenic on a moral level, since it applied veterinary methods to human beings. He mentioned some unnamed regions of Germany after the war, where because women outnumbered men, the idea was promoted that 'supernumerary' women should join freely with stronger and healthier men in order to fight demographic decline and to improve the race (*stirpis typus*).[28] Then he added that eugenics was right in its objectives, but not always in its methods. He found evidence for this statement in Harry Laughlin's proposal for the adoption of systematic sterilization measures in order to avoid the degeneration of race, as well as for the diffusion in the US of the laws for sterilization of the unfit, starting with the law passed in Indiana in 1907. Le Floch 'stressed that it was important that priests did not actively promote eugenics, because the theory faced the risk of being incorrectly interpreted and applied.' He also indicated that the clergy must promote the idea that the Catholic Church was the 'vera fautrix Eugenicae', the real advocate of eugenics. The conclusion of his report was centred on the fact that eugenics defined by Catholic morals aimed at being connected not only to physical health but also to the 'conjunction of the soul with god' and 'the subjects of the saints and servants of God (cives sanctorum atque domestici dei)'. Furthermore, according to Le Floch, marriage should not be prevented solely by illness, or some abnormalities, because it pertains to the law of nature; by getting married, even epileptics, alcoholics, syphilitics, and persons suffering from tuberculosis, committed, in some ways, a sin against charity, a fundamental precept for Catholic ethics. Charity demands strong adherence to the principle of not harming others and conceived as such, for Le Floch, it had an incomparable eugenic value. Le Floch concluded that books on this matter had to be written in Latin, not in vernacular language and therefore that the book by Mazzeo and de Giuseppe could not be approved '(approbari non potest)'.[29] The Holy Office subsequently accepted this statement.

The first time that the Holy Office had to face the question of eugenics it used mainly the arguments and disciplinary patterns elaborated in the second half of the nineteenth century when dealing with biopolitical issues and marriage. Eugenics thus emerged not as something questionable in its objective, but problematic mainly in its methods. Nevertheless, it is important to note that the first experts evaluation of eugenics produced within the Holy Office took into consideration both the experiences of the US sterilization laws and the German improvement practices. In his report on Mazzeo and de Giovanni's book, Le Floch mentioned explicitly the American eugenicist Harry H. Laughlin (1880–1943) and his proposals for the sterilization of the unfit.

This reference helps to explain some of the reasons why the encyclical *Casti connubii* would have discussed eugenics and sterilization. At the same time, the discussion within the Roman Inquisition about eugenics showed that this topic did not appear to be the main and absolute preoccupation of the Catholic Church. Eugenics was discussed together with the sexual education of young people, which appeared to be the main concern of this period. In some ways, the path leading to the first decree of the Roman Inquisition concerning eugenics in 1931 confirms this partial interest in eugenics and the main relevance of the question of sexual education, which was also the object of the encyclical 'Educatione iuventutis christianae' issued by Pope Pius XI on 31 December, 1929, exactly one year before *Casti connubii*. The first inquisitorial official view expressed on eugenics in the decree, issued on 18 March 1931 was reached from a discussion concerning both sexual education and eugenics. The decree presented two different decisions, one concerning sexual education, the other concerning the 'so called eugenics, or positive or negative' (sic dicta 'eugenica', sive 'positiva', sive 'negativa'). The decision of the inquisitors on eugenics was quite concise: 'it was to be considered as false and condemned following the statements of *Casti connubii*'.[30] This partial relevance bestowed on eugenics is confirmed by the same discussion that produced this decree. For the most part, the debate among the inquisitors had been focused on two main issues, the sexual education of children and young people and the teaching of the issues concerning the 6th and 9th commandment in theological seminars. Eugenics was the final argument of this debate, mainly taken and resolved through the simple reference to the encyclical *Casti connubii* statements.

A clear reference for understanding the causes of this debate did not emerge from internal discussions on eugenics. Some answers can be found in the origin of the encyclical. A history of *Casti connubii* has not been written yet and there have been some doubts about who wrote this document. According to Martine Sevegrand the author was the Belgian Jesuit Arthur Vermeersch, theologian first in Louvain and then at the Gregoriana University in Rome. John T. Noonan, who is one of main reference for the history of this encyclical, claimed that a large part of the text had been written by Vermeersch, while the German Jesuit Franz Hürth (1880–1953) wrote the part concerning sterilization.[31] According to recent studies, based on the documents of the archive of the Gregoriana University, Franz Hürth was the only author of the *Casti connubii*.[32] As a result, Vermeersch could be considered to have indirectly contributed to his direct relationship with Pope Pius XI as in the case of the Franciscan Agostino Gemelli (1878–1959), a former socialist and positivist, and then founder in 1921 of the Catholic University of Milan, and close advisor to Pope Pius XI for questions concerning, *inter alia*, biomedicine.

With regard to the discussion about the Latin translation of the encyclical, it clearly emerged that condemnation in *Casti connubii* of eugenics and even on sterilization was not unconditional. A few weeks after its publication, Franz Hürth informed Włodzimierz Ledóchowski (1866–1942), who was in charge of the Company of Jesus that there were differences about the judgment on sterilization between the last Latin version of the encyclical and the Italian and German versions. Concerning the morality of compulsory sterilization, the Latin version condemned the use of forced sterilization on innocents and people with hereditary defects, the German and Italian versions presented a complete condemnation of compulsory sterilization in itself.[33] The original version differentiated sterilization for criminal behaviour from eugenic sterilization and assumed the complete refusal only of eugenic sterilization. But in the published version this distinction disappeared. Hürth explained this distinction to Ledóchowski in order not to avoid contradicting German theologians and catholic authors who were in favour of forced sterilization for crimes. In order to correct the incongruity between the different versions of the encyclical that had been highlighted in newspapers, the journal *Acta Apostolicae Sedis* issued a short amendment of the Encyclical, restoring the original version.

An interesting position concerning eugenics can be found in the 1929 inquisitorial discussion on sperm culture as a diagnostic method to contrast venereal diseases. In the end, the Holy Office condemned this diagnostic procedure, using the theological principle 'non sunt facienda mala ut eveniant bona (you cannot do bad things in order to achieve good things)'. In the written version of his vote, the Secretary of the Holy Office, Rafael Merry del Val (1865–1930) former Secretary of State condemned sperm culture as a remedy even if it was the only effective way to treat gonorrhoea, because there was a right superior to that of the individual to heal his life. This right was

> the protection of moral and social life of individuals and peoples, and this right, or absolute necessity, is to maintain the Christian morality, the natural and divine law. Even when faced with medical

needs, we cannot sacrifice it on the altar of the Goddess eugenics. The man has a very different purpose than simply the physical preservation of the human race more or less accessible from medical care, moreover we treat of men and not of the breeding of animals.

On the same occasion, another important figure of the Catholic Hierarchy, the Secretary of State Pietro Gasparri (1852–1934) wrote a similar condemnation. He maintained that the principle 'non sunt facienda mala ut eveniant bona' was an absolute principle that prevailed over the therapeutic purpose; most of all in the case of the pursuit of 'a temporal good as the care of a sick body (bonum temporale, qualis est sanitas corporis infirmis)'.[34]

The Holy Office returned to eugenics a few years later in 1940, when it condemned euthanasia and again in 1941 when it condemned forced sterilization. It is not possible at this point in time to access the archival sources in the Vatican to be able to discover details of the debate among the members of Roman Inquisition about this issue, but it can be assumed that the interpretative perspective was the same as the one which had characterised the Catholic discipline of sexuality, reproduction and government of the body. This view stipulated that the order of nature cannot be modified by any intervention, for any motivation, and eugenics, considered to be good in terms of some of its goals, was considered unacceptable because of its methods, motivation and political agents, namely the state.

Notes

[1] See P. Weindling, *Health, Race and German Politics between National Unification and Nazism, 1870–1945* (Cambridge, 1989); W.H. Schneider, *Quality and Quantity: The Quest for Biological Regeneration in Twentieth Century France* (Cambridge, 1990); F. Cassata, *Building a New Man. Eugenics, Racial Science and Genetics in Twentieth Century Italy* (Budapest and New York 2011); S. Leon, *An Image of God. The Catholic Struggle with Eugenics* (Chicago and London, 2013); M. Turda and A. Gillette, *Latin eugenics in Comparative Perspective* (London, 2014).

[2] See D. J. Dietrich, 'Catholic Eugenics in Germany, 1920–1945: Hermann Muckermann, S.J., and Joseph Mayer', *Journal of Church and State*, 34/ 3 (1992), 575–600; R. Graham, 'The Right to Kill in the Third Reich: Prelude to Genocide', *The Catholic Historical Review*, 62/1 (1976), 56–76. On Muckermann's role, see H.W. Schmuhl, *The Kaiser Wilhelm Institute for Anthropology, Human Heredity and Eugenics, 1927–1945. Crossing Boundaries* (Dordrecht 2008), pp. 97–106.

3. See S. M. Leon, '"Hopelessly Entangled in Nordic Pre-suppositions": Catholic Participation in the American Eugenics Society in the 1920s', *Journal of the History of Medicine and Allied Sciences*, 59/1 (2004), 3–49; Leon, *An Image of God*; C. Rosen, *Preaching Eugenics: Religious Leaders and the American Eugenics Movement* (Oxford, 2004); A.H. Marouf, jr, *The Eugenics in Anglo-American Thought* (Athens and London, 1996), in particular chapter 5 'Catholic interpretations of Eugenics Rhetoric'.

4. E. Lepicard, 'Eugenics and Roman Catholicism. An Encyclical Letter in Context: Casti connubii, December 31, 1930', *Science in Context*, 11/ 3–4 (1998), 527–44.

5. Founded in 1542 the Sacred, Roman and Universal Inquisition of the Holy Office was the most important congregation in the Catholic Church, focused on the defense of the orthodoxy of Catholicism. It had a superior jurisdiction among the other Congregations and its decisions had a doctrinal value. During the Second Vatican Council in 1965, the Congregation adopted the name of Congregation for the Doctrine of Faith.

6. For the use of the concept of discipline, see M. Foucault, *The History of Sexuality*, Vol. 1, *The Will to Knowledge*, (London, 1998) [originally published in 1976].

7. A. Desmazières, *L'inconscient au paradis. Comment les catholiques ont reçu la psychanalyse (1920-1965)* (Paris, 2011).

8. M. Artigas, T. F. Glick, R. A. Martínez, *Negotiating Darwin. The Vatican Confronts Evolution, 1877-1902* (Baltimore, MD, 2006).

9. See J. T. Noonan, *Contraception: A History of its Treatment by the Catholic Theologians and Canonists* (Cambridge Mass., 1965). Recently two books have explained the concrete political relevance of *Casti connubii* into contemporary debates concerning reproduction. Connelly has pointed out that Catholic Church organized the opposition to the development of birth control politics by the international institutions, as the UN, FAO, UNESCO, following the statements of the encyclical of 1930; see M. Connelly, *Fatal Misconception. The Struggle to Control World Population*, (Cambridge, Mass., 2008). Thomas Banchoff, on the other hand, has shown the guiding role of *Casti connubii* for the opposition movements towards embryo politics in France, the UK, the US and Germany, after the birth of Louise Brown in 1978. See T. Banchoff, *Embryo Politics. Ethics and Policy in Atlantic Democracies* (Ithaca, 2011).

10. The various drafts of the encyclical can be found in G. Antoniazzi, ed., *L'enciclica Rerum novarum. Testo autentico e redazioni preparatorie dai documenti originali* (Rome, 1957). For the discussion of *onanism* and birth control in the Catholic Church, see C. Langlois, *Le crime d'Onan. Le discours catholique sur la limitation des naissances (1816-1939)* (Paris, 2005) and E. Betta, '«De usu imperfecto matrimonii». Il Sant'Uffizio e il controllo delle nascite', *Quaderni storici*, 1 (2014), 141–82.

11. The vote presented 193 in favour, 67 contraries and 40 abstained. The final resolution of the conference admitted the use of non-natural birth control methods only in the case of necessary legitimate reason, first of all the woman's health. See G. Cuchet, 'Quelque données concernant l'encyclique Casti connubii', in J. Prévotat (ed.), *Pie XI et la France* (Rome, 2010), pp. 347–67.

12. *Penitenzieria* intervened in 1816, 1822, 1823, 1842, 1847, 1853, 1876, 1880, 1886, and then in 1901, 1904, 1916 (two times); *Holy Office* in 1851, 1853, 1897 and then in 1922 and 1929. See Noonan, *Contraception*. For a list of intervention with text of questions and answers, see P. Harmann Batzill, *Decisiones sanctae sedis de usu et abusu matrimonii* (Torino, 1937).

13. *Casti Connubii: Encyclical of Pope Pius XI on Christian Marriage to the Venerable Brethren, Patriarchs, Primates, Archbishops, Bishops, and other Local Ordinaries Enjoying Peace and Communion with the Apostolic See*, 1930, December 30. Quotations are at 29 and 69. The

version of the encyclical in H. J. Denzinger, *Enchiriodion symbolorum definitionum et declarationum de rebus fidei et morum* (Bologna, 2009), or on the Vatican Web Site: *https://w2.vatican.va/content/pius-xi/en/encyclicals/documents/hf_p-xi_enc_19301231_casti-connubii.html*.

[14] *Casti connubii*, p. 126.

[15] For the details of the case see E. Betta, '1873: la contraccezione all'Indice', in V. Lavenia, G. Paolini (eds), *Riti di passaggio, storie di giustizia. Per Adriano Prosperi*, vol. III (Pisa, 2011), pp. 35–42.

[16] Concerning this passage of Augustin see Noonan, *Contraception*, pp. 427–8.

[17] *Casti connubii*, p. 66.

[18] *Casti connubii*, p. 68.

[19] *Casti connubii*, p. 70.

[20] For the whole reconstruction and analysis of this normative process see E. Betta, *Animare la vita. Disciplina della nascita tra medicina e morale nell'Ottocento* (Bologna, 2006).

[21] The history of artificial insemination and the judgment of the Catholic Church in E. Betta, *L'altra genesi. Storia della fecondazione artificiale* (Roma, 2012).

[22] Porro improved the so-called caesarean section with the excision of the uterus and the adnexae when he was professor of Obstetric at the Pavia University in the volume *Della amputazione utero-ovarica come complemento di taglio cesareo* published in Milan in 1876, fifteen years after the foundation of the Kingdom of Italy. He asked to Parrocchi for advice only as a catholic, not because he needed it, nor it required any approval from the Holy Office, contrary to what sustained in J. G. Ryan, 'The Chapel and the Operating Room: The Struggle of Roman Catholic Clergy, Physicians, and Believers with the Dilemmas of Obstetric Surgery, 1800–1900', *Bulletin of the History of Medicine*, 76/3 (2002), 461–94, esp. 472.

[23] On this case, see L. Mendes, R. Ferreira Vieira, 'O "caso Abel Parente", os homens de letras e a disseminação do saber científico nos primórdios da República', *Revista Maracanan*, 13 (2015), 127–45.

[24] F. De Castro, *O invento Abel Parente no ponto de vista do directo criminal da moral publica e da medicina clinica* (Rio de Janeiro, 1893).

[25] G. Moscati, 'Prefazione', in G. de Giovanni, and M. Mazzeo, *L'eugenica* (Naples, 1924), pp i–iii.

[26] See J. Creusen, 'Father Arthur Vermeersch, S. J., 1858–1936', *Studies: An Irish Quarterly Review*, 26/103 (1937), 429–38.

[27] The copy of the file in the archive indicates February 27, 1925 as the registration date of the discussion. Cf. Archive of the Congregation for the doctrine of Faith, *Rerum variarum*, 1931 24, vol. 1.

[28] H. Le Floch, 'De sic dicta "Educatione puritatis". Votum', ACDF, *Rerum variarum*, 24/1 (1931), f. 11, p. 70.

[29] Le Floch, 'De sic dicta "Educatione puritatis"', p. 75.

[30] *De educatio puritatis et eugenica*, in Archive of the Congregation for the doctrine of Faith, *Dubia Varia*, 1931, p. 165.

[31] M. Sevegrand, *Les enfants du bon dieu. Les catholiques français et la procréation au XX siècle* (Paris, 1995).

[32] L. Pozzi, 'Chiesa cattolica e sessualità coniugale: l'enciclica Casti connubii', *Contemporanea. Rivista di storia dell'800 e del 900*, 3 (2014), 387–412.

[33] For the details of this case, see L. Pozzi, 'L'enciclica *Casti connubii*, l'eugenetica e la sterilizzazione forzata', *Römische Quartalschrift für christische Altertumskunde und Kirchengeschichte*, 2 (2014), 226–39.

[34] For the quotation and the case, see Betta, *L'altra genesi*, pp. 144–9.

EUGENICS, SEX REFORM, RELIGION AND ANARCHISM IN PORTUGAL[1]

Richard Cleminson

Introduction

Although the interconnections between Catholicism and eugenics were multifaceted and were textured in different ways according to the context in which they arose, in the Portuguese case, it was perhaps the searing debate in the early 1930s between Jaime Brasil (1896–1966), a vociferous proponent of sexual science and eugenics, and the Catholic establishment that showed most clearly how conflictive this relation could be. In mid-1932, the journalist, anarchist activist and sex reformer Jaime Brasil published his *A Questão Sexual*, a highly detailed volume of some 480 pages covering most aspects of sexual expression.[2] Faithful to a common format in European works on the 'sexual question', the book dealt with 'morbid' and 'normal' sexuality.[3] Within the section on 'normal' sexuality there were chapters on 'Natal Matters' and 'The Selection of the Species', covering inheritance and eugenics[4] – areas that trespassed on what Catholics believed to be their terrain of reference. The vehement reaction from the Jesuit review *Novidades* came quickly over the summer of 1932, in the form of several articles classifying Brasil's book as communist propaganda and as a source of moral corruption of the youth. In turn, Brasil responded to the Catholic campaign, denouncing the 'hateful vomit of the padres' in a short book that reproduced and commented on the increasingly charged texts that were exchanged between the two camps.[5]

In *A Questão Sexual* Brasil put forward arguments on the benefits of population limitation and 'conscious procreation', and stated that his book was aimed at dignifying women's lives and making contraceptive methods available as part of his overall objective of clarifying 'certain aspects of Existence, which are surrounded everywhere by ridiculous preconceptions, harmful practices and magical taboos'.[6] It is important to note, from the beginning of this article, that eugenics was articulated by Brasil as one of a set of resources allowing him to address the 'sexual question', and should be seen as being complementary to this discourse on sexuality and not necessarily as the main point of conflict with the Catholic Church. *A Questão Sexual* should be understood to form part of what was the broader contemporary endeavour of 'sex reform' articulated by 'progressive' sectors of society across Europe. The condemnation by *Novidades* was a general one and did not focus particularly on eugenics but reflected the Catholic Church's indignation at the temerity of a secular author who attempted to place sexuality within a scientific and not a religious moral framework. The confrontation was, essentially, a struggle for control over who could legitimately and authoritatively discuss issues of sexuality. It will nevertheless be argued here that, in order to understand this conflict, we do need to set Brasil's *A Questão Sexual* within the context of both his own writings in the early 1930s on eugenics and sexuality, and the relations between the two, as well as within the broader context of Catholic discourse on eugenics and sexuality. In addition, the sociopolitical context of the early 1930s with respect to both the anarchist movement and the consolidation of the Salazar dictatorship in Portugal are essential elements in this debate. By taking all these factors into account, we can perceive how eugenics came to be a mobile resource for Brasil, enabling him to reinforce the scientific dimensions of his work on sex reform; we can also see how eugenics and sexuality became bound together in both Catholic and anarchist mentalities. Portuguese Catholicism will be seen to have been particularly reactionary with respect to issues of sexuality and eugenics. Other countries' Catholic Churches reacted differently.[7]

If the relationship between eugenics and religion is complex and problematic, the added ingredient of the reception of eugenics within different strands of the labour movement makes it even more so.[8] In many cases, and notably in 'Latin' countries, labour movements rejected religious control of issues relating to sexuality, reproduction and family morality. They often rejected eugenics, too, if they came across it. However, there were movements or sectors of movements that, while rejecting religious 'interference' in sexual questions, accepted some of the postulates of eugenics. This acceptance responded to two main complementary motivations. Religion was rejected as a legitimate source of explanations of the world and as a guide for social and sexual relations.

In contrast, scientific thought was accepted and promoted, especially in the form of theories of evolution that were read positively to provide apparently progressive interpretations of human existence, social evolution, women's roles and notions of freedom.[9]

Not only was Jaime Brasil outside the mainstream eugenics movement in Portugal;[10] he spoke and wrote from a particular perspective within the Portuguese and international anarchist movement that had developed an interest in sexual matters from the late nineteenth century onwards. Although eugenics was a minority interest in Portuguese anarchist circles, questions relating to the role of women in society, sexual hygiene and contraception (understood as 'neo-Malthusianism'), were durable areas of interest within anarchist circles from the late nineteenth century until well into the 1940s.[11] The fact that Brasil moved effortlessly from sexual hygiene to eugenics in A Questão Sexual and, as we will see, in his other writings, is testimony to this interest and to the association, common in the French and Spanish anarchist movements among others, between sex reform, birth control and eugenics.

This article, as a secondary objective, therefore intends to make a contribution to the history of the reception of eugenics within Portuguese and international anarchism. The debacle over Brasil's work took place as the fascistic Estado Novo under Dr António de Oliveira Salazar was being consolidated – a process that had begun when the democratic Republic, established in 1910, was overthrown by a military coup in 1926 and many basic liberties, including that of the freedom of the press, were placed under threat or had already been severely compromised. The anarchist movement operated clandestinely by 1932 and was, the following year, along with other oppositional groups, formally illegalised. Despite this, the structures of the movement, and particularly its syndicalist tendency encapsulated by the CGT (General Confederation of Labour) were maintained up to and beyond the revolutionary attempt to overthrow Salazar on 18 January 1934 when repression severely curtailed the viability of the movement.[12] Despite this, anarchist propaganda was still produced up the end of the 1950s.

Theories of heredity, progress and religion

Following the work undertaken on the relationship between anarchism, the spread of theories of heredity and progress and the rejection of religion in countries such as Spain and France,[13] it can be posited that certain sectors of anarchist movements based their (limited) adoption of eugenics on their approval of science as potentially liberatory if placed in the 'right' hands – that is, far from the influence of the bourgeoisie and institutionalised religion. Such an acceptance of scientific ideas in general, and understandings of human, animal and plant evolution in particular, would be articulated alongside diffuse notions of inheritance, which vacillated between what might be termed 'Mendelian' and 'Lamarckian' understandings (without usually being referred to as such by anarchists). The acceptance of eugenics, in some cases constructed on the postulates of neo-Malthusianism, conjoined with birth-control propaganda, anti-militarism and 'rational' non-religious education, provided a scientifically respectable means by which the anarchist desire for social and moral perfection – understood as the elimination of hierarchies and the establishment of a stateless society with no centralised power – could find a home. I have argued elsewhere that, in the Spanish case, these disparate strands came together and allowed for a platform upon which an anarchist expression of eugenics could sit.[14] In the Portuguese case, many of these ingredients were present,[15] but the connection between them and eugenics was rarely forged and remained limited to a few figures such as Jaime Brasil. The full reasons for this lack of engagement cannot be entered into here, but some brief remarks on the context of ideas on Malthus, heredity, and sexual matters within Portuguese anarchism are required.

Alongside Darwin, the means by which scientific discussion on species change (eventually 'evolution') entered Portugal, in common with many other 'Latin' countries, was via discussions on inheritance and transformism in the context of Lamarckism.[16] One of the means by which eugenics made its entry into the Portuguese scientific milieu was in a discussion of Lamarckism by University of Oporto scholar Americo Pires de Lima (1886–1966), who wrote his *A evolução do transformismo* in 1912.[17] Pires de Lima, defending his thesis in late 1912, praised the work of Charles Darwin, Jean-Baptiste Lamarck, Hugo de Vries, August Weismann and Herbert Spencer in order to explain the mechanism of inheritance. In a late addition to his work, as an appendix just before reading his thesis, he assessed the 'new' thought of Gregor Mendel.[18] Mendel's theory was largely rejected by Pires de Lima as undemonstrated and, despite having been employed to bolster eugenic ideas at the 1912 Eugenics Congress in London, was taken to undermine the case for eugenics precisely because its claims were undemonstrated. Any resultant prohibition of the marriage and reproduction of certain individuals based on insights drawn from Mendelism were cast by Pires de Lima as 'inefficacious, [. . .] an affront and unjust'.[19]

Reflections on Malthus's warnings on population growth and Darwinian explanations of the struggle for existence began to percolate and be debated in the Portuguese anarchist movement in the early twentieth century. An early exponent of the neo-Malthusian interpretation of Malthus's doctrine on population imbalance can be found in numerous figures connected to the anarchist movement, such as Ângelo Vaz (1879–1962) and in reviews such as *Paz e Liberdade*, subtitled 'A Monthly Anti-Militarist, Anti-Patriotic, Revolutionary Syndicalist and Neo-Malthusian Review'.[20] Some authors, particularly when discussing the ideas of the French educational and sex reformer Paul Robin (1837–1912), the inspiration behind the International League for Human Regeneration, articulated the need to improve the 'quality' of the population but did not do so within the discursive frame of eugenics.[21]

As Ana Leonor Pereira has argued, in Portugal, one of the most significant early interventions in debates on social progress, the usages of notions of the 'struggle for existence', and ideas on the mechanisms of inheritance was the 1910 libertarian work by João Evangelista Campos Lima (1887–1956).[22] The first part of his book, an analysis of the thought of Cesare Lombroso, Malthus and Darwin, among other thinkers, was followed by an overview of the state of the workers' movement in Portugal to date. A strong environmentalist thesis on the criminogenic and unhealthy nature of capitalist society was advanced. Lima wrote: 'Let the environment be transformed, let all be given the necessary means of existence and, fear not, the biological law will falter: degeneration will be limited and, even though it will subsist some time afterwards because of the residues that heredity transmits, it will eventually disappear'.[23] The so-called 'struggle for existence' was seen as a ruling-class ploy to aid its own survival and dominance. Lima reaffirmed this line of reasoning in the 1920s when he argued that the weak and the degenerate, as they normally attracted one another and as their unions were usually infertile, would die out in the future.[24] He argued that the lack of naturalness of the marriage convention, which allowed for the weakest to thrive, would not survive into the future and that the fixation of the 'superior qualities of individuals and the elimination of the weakest by means of the lack of fecundity of their sexual relations' would be the result.[25] Such a process would guarantee the progressive perfectibility of humankind.[26] Jaime Brasil would continue this kind of analysis into the 1930s.

Jaime Brasil: From 'Conscious procreation' to eugenics

Apart from his collaboration with the more trade-union oriented *A Batalha*, Jaime Brasil was a significant figure in a variety of other publications. One of these was the already mentioned *O Globo*, a critical cultural review, which, despite the dictatorship, maintained a weekly periodicity under Brasil's editorship and directorship from January 1930 up to the end of July of the same year. An eclectic mix of articles was published in the review, which enjoyed high-quality presentation and eye-catching imagery. Articles on the women's movement, reproduction, morality, film, emigration and nudism were just some of the themes approached. In May 1930, Brasil wrote on the subject of 'voluntary procreation' and argued in favour of a doctrine that he declared would be termed 'biosofia'.[27] The idea of voluntary procreation for women – by that date common currency in the French and Spanish anarchist movements – argued in favour of women having children when they desired and called for women to 'impede the reproduction of people whose physical or moral taints, where scientifically identified, prevent them from perpetuating the species' as a 'duty' towards society.[28] In this process, the individual should be subordinated to the species. Pre-matrimonial certificates were insufficient, and divorce was not adequate either for the selection of a 'good human product'. Instead, preventive neo-Malthusian measures should be extended and doctors should participate in a new institute that would be 'in charge of overseeing the conservation and perfecting of the species'.[29] This body would also prevent certain individuals from reproducing and would provide them with the means to avoid doing so. Such a questionable institution in libertarian terms would also allow for or encourage the voluntary castration of alcoholics, syphilitics and madmen – an intervention that would supposedly regenerate the species. Brasil named the French authorities Adolphe Pinard and Alfred Binet to back up such ideas.

It was not long before Jaime Brasil began to articulate such notions explicitly as part of a programme of eugenics. Although his engagement with eugenics was significantly more extensive in his later *A Questão Sexual* and his *A Procriação Voluntária* of 1933,[30] in the slim *O Problema Sexual* (1931) Brasil touched on the international eugenics movement just a few years after the 'founding' document of Portuguese eugenics, the report written by the Oporto anthropologist A. A. Mendes Correia, *O problema eugénico em Portugal* (1927).³¹ Brasil's *O Problema Sexual*, as the title suggests, was devoted principally to the need for sex education. The work began with a number of assertions that were commonplace in the transnational anarchist movement: that humans ('man' in the original) had been born 'good and free' but had later been enslaved and had given themselves over to hatred. Only through the use of intelligence and knowledge could they recoup their lost freedoms and goodness. But how, Brasil asked, could this be guaranteed if vices and bad traits existed in humanity?³² These disadvantages, he responded, could be eliminated by means of integral education, which would in turn be based on morality and the truths of biology. Given the fact that sex was at the root of life, Brasil continued, it was to be lamented that it was hardly treated objectively at all; the sexual organs were taboo matters in schools and in life in general.³³ Brasil marshalled a number of voices to register his disapproval of this situation including some Catholic figures such as the Archbishop of Paris, Cardinal Verdier, who spoke to this effect at the VIIth National Congress on Christian Marriage held in Paris.³⁴ Brasil even suggested the creation of an Institute of Sexuality and Eugenics to take care of the teaching of these matters to the population at large.³⁵

This Institute and other initiatives would provide boys and girls with information on the dangers of prostitution, syphilis and other venereal diseases, the advantages of premarital medical examinations and the role of voluntary and conscious maternity, and would establish clinics to dispense 'eugenic devices' (probably contraception).³⁶ In the chapter on the issue of voluntary procreation, Brasil praised the work of the recently established World League for Sex Reform (WLSR), at whose congress (Geneva 1927) the question of sex education had been discussed.³⁷ Voluntary procreation was described as one of the fundamental rights of women.³⁸ It would appear, however, that neither the Institute – at least in the form that Brasil envisaged – nor the Portuguese chapter of the WLSR came into existence.

Brasil soon moved on to address the question of impeding procreation in those cases where poor traits were deemed hereditary and incurable. These should be taken cognisance of in order to prevent procreation. Brasil argued that the transmission of degenerative traits, such as tuberculous conditions, had been demonstrated scientifically and, as a consequence, in certain American states 'castration' of delinquents had been permitted.³⁹ Although such a step was not explicitly advocated or condemned by Brasil, his view was that a more rational and human approach rather than 'these mutilations or privations' would be what he called the prophylaxis of conception.⁴⁰ It was this method that the neo-Malthusians Eugène and Jeanne Humbert had defended at the 4th congress of the WLSR, held in Vienna in 1930.⁴¹ Even the Lambeth Conference had reversed a decision previously taken in 1920 to disapprove of contraception, by allowing it if its use was 'moral'.⁴² Such a move, of course, provoked the papal encyclical *Casti connubii* criticising those who had strayed from the Christian path, and it was these sectors of the Church that Brasil condemned strongly in his final section.

Brasil finished his discussion of voluntary procreation by praising what he identified as the recent upsurge – a 'silent revolution', in Norway and elsewhere – whereby free unions outside marriage had increasingly become the vogue.⁴³ This oft-cited anarchist remedy for the authoritarian nature of marriage was classified by the author not as a utopia but as a strident reality. Finally, eugenics and its supporters, guided by the recent progress in genetics, would provide a 'perfect humanity' in the future.⁴⁴ Such a scenario could be halted only by a generic 'eles' ('they') who would attempt to scupper emancipatory developments. These sectors wanted everything to remain the same; they were the 'frightful forces of reaction in the service of a depraved and greedy capitalism' incarnated in a form of 'corrupt and ignorant clericalism that, incapable of discussing ideas, denounces its adversaries to the police'.⁴⁵

This short book by Brasil combined explicit praise for scientific developments, a libertarian political stance in favour of sex education and women's rights with a condemnation of the most reactionary sectors of Catholicism and capitalism. It drew on a battery of sex reformers to support its cause, including Iwan Bloch, Havelock Ellis, Sigmund Freud, Egas Moniz and Pierre Vachet, and combined them with prominent anarchist thinkers from across Europe (Charles Albert, E. Armand and Élisée Reclus).⁴⁶ The references to eugenics per se were not extensive but the understanding of voluntary procreation as part of a process leading to the perfection of human types, the discussion of hereditary traits and 'castration' (i.e. 'eugenic sterilisation') and suggestions for less authoritarian alternatives place *O Problema Sexual* within the anarchist tradition of eugenics and, more significantly, within a tendency whereby discourse on conscious procreation and neo-Malthusianism were beginning to take on explicitly eugenic overtones. The added dimension interrogating the various stances of the Christian Churches, together with the condemnation of those that Brasil took to be the most reactionary sectors, also confirm the piece's presence within this tradition. The work also, quite evidently, set the stage for future conflict with the Catholic Church.

A far more explicit account of the science of eugenics was to be elaborated in Brasil's *A Questão Sexual* a year later. His thought on the subject was evidently undergoing rapid evolution, as it took on greater sophistication and showed its debts to broader international eugenic discourses.[47] After having argued in favour of contraception as an economic and moral necessity, Brasil justified such means as being devoted principally to the selection of the species. Women should be free to have children when they wished and should be able to satisfy their own sexual desires without the risk of pregnancy.[48] As part of the 'selection' process, women should be able to have sexual relations with men until finding the right partner in order to have a child. This was an individual right, he declared.

The collective should, however, also be considered. Under the capitalist system (which, Brasil averred, would not last for much longer) it was necessary for the proletariat to cease reproducing in order to limit the growth of surplus workers, reduce poverty and unemployment and combat wage reductions.[49] Such assertions placed Brasil firmly in the anarchist neo-Malthusian camp, as expressed in reviews such as *Salud y Fuerza* (1904–14), which had argued for a similar strategy.[50] Beyond these measures, however, it was necessary to 'avoid procreation between individuals who are ill and tainted'. Such a concern had been discussed in earlier evaluations of eugenics in Portugal, for example, in the writings of the psychiatrist Miguel Bombarda (1851–1910) in 1910 and later on among eugenicists in the 1920s, and, in this sense, Brasil's referral to this debate was not exceptional for the period.[51] But how did it fit with a doctrine supposedly devoted to the liberation of individuals and opposed to authoritarian methods?

It was the relationship between individual rights and the collective health of the species that was at the forefront of Brasil's rationalisation of this problem. The limitation of reproductivity in these individuals was, to Brasil's eye, to their own benefit, but it would be 'cruel and absurd to prevent them from having sexual relations'. Their own natural libido demanded the exercising of a normal sexual life but without the burden of offspring, which, even though it was not stated, was something that would be possible by means of the use of contraception.[52] This would allow for 'the healthy, apt, strong, intelligent' types to procreate.[53] The programme of eugenics was now not just an ideal but a 'laboratory science', the prescriptions and discoveries of which should be obeyed.⁵⁴ Drawing on the French eugenicist Charles Richet's work, *Sélection humaine* (1919), Brasil repeated his support for conscious procreation and scientific developments as routes towards human perfection.⁵⁵

Richet (1850–1935) was the first of numerous international commentators on eugenics to be referred to by Brasil. Others included Marie Stopes (1880–1958) and Leonard Darwin (1850–1943), President of the British Eugenics Society. Darwin was quoted in order to illustrate the two main tendencies within eugenics – positive and negative – and to provide ballast for the argument to promote the reproduction of those with 'superior faculties'.⁵⁶ Although Brasil's engagement here with eugenics was more extensive than in previous publications, it was still of a somewhat popularised nature. Darwin, for example, was referred to but the precise reference was not supplied. In the final section of the book, a whole range of measures were proposed in order to provide guidance on how sexuality would be in the future. The 'sexual revolution' that was under way would draw on conscious procreation, nudism, the elimination of prostitution, opposition to masturbation, and different models of cohabitation including monogamous and polygamous relationships. This revolutionary change was deemed to be inseparable from the broader social revolution.⁵⁷

With *A Questão Sexual* Brasil admitted that he had not wanted to write a doctrinal, normative, professorial or didactic work. Instead, he had aimed to educate his readers about sexuality through knowledge and understanding.⁵⁸ Such an aim, of course, coincided with his anarchist and indeed journalistic preferences, and it continued into his *A Procriação Voluntária*, the volume that appeared after his counter-attack on *Novidades*. Before considering the central debates contained in this riposte, we will go forward in time to analyse *A Procriação Voluntária* for its continuation of Brasil's eugenics. The final part of this article will return to *Os Padres e 'A Questão Sexual'* and the conflict between Brasil's anarchistic sex reform project and eugenics and Catholicism.

'Systematic and progressive perfection' through voluntary procreation

The relationship between the individual and the collective, constant in the anarchist political repertoire, was revisited by Brasil in his *A*

Procriação Voluntária. The mechanism by which the balance would be achieved was through the new science of eugenics. Eugenics would engender a process of systematic and progressive improvement in the individual. Its methods were praised by Brasil as the route to what philosophical schools, religious thought, scientific discoveries and social systems had yearned after for centuries: the attainment of human perfection. As the preface to *A Procriação Voluntária* proves, eugenics had gained centre stage in Brasil's thought by 1933. In light of his suggestion that the study of the economic ideas of neo-Malthusianism needed to be combined with the scientific premises of eugenics, Brasil, like many other anarchists, sealed the association between the two in anarchist ideology.[59] *A Procriação Voluntária* was composed of two parts, the first being a theoretical exposition on the regulation of birth, imbued with eugenic concepts, and the second dedicated to practical birth control techniques faithful to the subtitle of the work, 'The Means of Avoiding Pregnancy'.

In the theoretical section, Brasil appeared to be well informed about the debates and about the relevant thinkers. Havelock Ellis, Francis Galton, J. Huxley, H. F. Osborn, Charles Davenport and Renato Kehl were all referred to, along with their relevant institutes and publications. This first part was also broad, and Brasil did not limit itself to discussions on eugenics alone. Chapters discussed the principles of eugenics, the basis of neo-Malthusianism, the birth control movement, the liberation of love and sexual prophylaxis. While it must be emphasised that all these sections were considered by Brasil to be interrelated and complementary as part of an overall whole, it is the section specifically on eugenics that will be discussed here. The various elements were, nevertheless, subject to considerable repositioning, and one has only to think of the discussion in a Spanish anarchist review of neo-Malthusianism and eugenics as complementary but different theories by the Spanish educationalist Luis Huerta in 1930.[60]

Brasil gave an account of the history of eugenics from its Galtonian starting point – reference is made to Galton's *Inquiries into Human Faculty* (1883) as the text that first mentioned 'eugenics' – and recounted its intellectual forerunners, including Darwin's debt to Malthus with respect to the idea of the struggle for existence (in fact, the debt of Darwin to Malthus's idea on population pressure).[61] The history and current state of the international eugenics movement was also recorded, and brief remarks were made on the French, Romanian, Italian, Russian, Estonian, North American and British movements. Few comments were made on the Portuguese counterpart, save the remark that the Brazilian Renato Kehl had given a talk to the Portuguese Anthropological Society and the aforementioned note on Almerindo Lessa's potential involvement in the WLSR.[62]

Returning to the thought of Leonard Darwin, Brasil explained that there were two principal types of eugenics: positive and negative. The positive form needed little explication: 'Any couple of healthy individuals, living under normal economic conditions, needs no advice on procreation within reasonable limits.'[63] This 'normal' couple would be composed of individuals who were 'healthy, with no ancestral taints, strong, young, intelligent and educated in matters pertaining to sexuality'.[64] The biological basis of eugenics was considered by Brasil to derive from similar techniques used for animal rearing ('zootecnia') in order to select races and improve types.[65] What, however, was to be done with the 'inferior' types? Brasil was quite clear on this. The healthy had to support the ill and the 'sub-products of humanity' in hospitals, asylums and through welfare schemes, and it was the task of negative eugenics not to eliminate these but to prevent them from being born in the first place.[66] In addition, certain states or conditions, following Marie Stopes's book on contraception, were deemed 'counter-indicative' for reproduction.[67] These included hereditary blindness, syphilis, tuberculosis, epilepsy, diabetes, mental disorders and, in women, the malformation of the pelvis.[68] Some importance was given to the environmental causes of degeneration and poor traits, with heredity seen to provide some 50 per cent of character; 15 per cent of character came from education, 25 per cent from the environment and 10 per cent from one's physiological state.[69] Such a formula was viewed as too rigid by Brasil, but worth taking into account as the neo-Malthusian Manuel Devaldès (1875–1956) had argued in his book on conscious maternity.[70] Devaldès was a favourite author among the anarchists in France, Portugal and Spain for his plain-speaking approval of women's right to sexual pleasure without reproduction and for his open stance on sexual education, and it was his work that Brasil used to frame the preface of his *A Procriação Voluntária*.[71]

The 'inheritance formula' mentioned by Brasil was indicative of discussions at the time on the importance of heredity with respect to environment and the ongoing deliberations in the eugenics movement itself. Those anarchists in Portugal, such as Campos Lima and Brasil, who raised this issue oscillated between an almost complete faith in the environment to improve humanity through to some kind of combination of environmental and hereditary influences. Brasil himself mentioned that certain diseases were transmitted through uterine means, tuberculosis made offspring more 'predisposed' to disease and certain illnesses were deemed to be hereditary. Such a concession to vague notions of predisposition, uterine inheritance and the environment was typical, first, of anarchist thought on the subject and, second, of many eugenicists who operated in the 'Latin' eugenics tradition.[72]

A number of ambivalences can readily be detected in the text. Brasil's talk of 'sub-humans', 'inferior types' and 'criminals' fitted, with difficulty, with an anarchist model of egalitarianism and the idea that capitalism was at the root of poverty and crime. In the last section of his chapter on eugenics, Brasil seemed to suggest that it was legitimate to prevent such types from reproducing but, on the other hand, he wrote that it was 'too cruel to try to prevent the poor from the pleasures that reproduction may bring them'.[73] Indeed, the eugenic strategy for the rich and the poor, reflecting a class analysis, would be different. The poor could employ methods allowing for a 'conscious regularization of birth' with the 'absolute suppression [of births] for the poorest' and this would allow for economic wealth for all and the happiness of perfection for humanity.[74] For the rich, a different outcome was suggested. The rich, he stated bluntly, 'had no right to exist'. Wealth, Brasil argued, was a 'taint that deforms character' and was rooted in the 'anal character' that psychoanalysis studied. The 'caste of the rich', therefore, carried its own seeds of destruction and would soon 'disappear from the surface of the earth'.[75]

It may be assumed that the methods to be employed for the 'absolute suppression' of births in the poor would be contraceptives. But in the second part of the book Brasil discussed other practical birth control techniques, and these included sterilisation. In a somewhat detached set of observations, Brasil noted that sterilisation was a process that had been disseminated as a solution in many parts of the world – as had vasectomy, a technique that had been persecuted. In England, sterilisation was proposed by some and opposed as irreversible by other eugenicists. In America, the Human Betterment Foundation approved of sterilisation and X-rays could be employed for such a purpose.[76] It would appear, therefore, that although Brasil advanced contraceptive methods as the key to impeding the reproduction of the 'unfit' and for the poor to control their own natality, sterilisation was clearly not rejected as a means towards the same end. This was unusual for anarchist commentators on the question of the legitimacy of sterilisation.[77]

Novidades and the Catholic response to *A Questão Sexual*

The relationship between Catholicism and the Salazar regime was not straightforward, and although Salazar was broadly in favour of Catholic moral precepts he was wary of the power of the Church regarding the day-to-day operation of the regime, its role in education and the sociopolitical influence that it could accrue.[78] But the regime adopted as its slogan, 'God, Nation and Family', and this triple liaison effectively sealed the influence of the Church in the regime's affairs and development. Traditional Catholicism was a 'crucial contributor' and 'determining influence' on the regime, and affected the wording of the constitution and the declaration of the regime's principles,[79] thus expressing a 'common ideological and political nucleus that was corporatist, anti-liberal and anti-communist'.[80] The daily paper *Novidades* grew out of a late nineteenth-century concern to provide intellectual and moral guidance to the population from a more 'neutral' position, and the journal was established in 1885 under the wing of the Progressive Party politician Emídio Navarro and the army officer and journalist Eduardo Noronha.[81] It became one of the principal Catholic publications of the twentieth century in Portugal and, having gone through various publishing periods, was still produced as a daily by the time of the debacle with Jaime Brasil. *Novidades* was not simply a reactionary publication, however; it condemned fascism and Mussolini in particular as examples of 'pagan' politics whereby the state claimed all authority (over and above the Church),[82] and Hitler's party and the Nazi regime also came in for much criticism throughout the early 1930s for their nationalism, racism and violence.[83] When Salazar addressed the first congress of the regime party, the União Nacional, and criticised the 'pagan totalitarianism' of fascism in 1934, it met with the resounding approval of *Novidades*.[84] Despite evident common causes, however, the eventual text of the Concordat signed in Rome between the *Estado Novo* and the Church on 7 May 1940 maintained the independence of the state vis-à-vis the Church. Certain hangovers from the anticlerical Republic were maintained such as divorce for those married during the republican period and for non-Catholics.[85] Catholic opposition to the regime and those forces broadly under the banner of 'social Catholicism' were kept in control by Salazar and the Church until the late 1950s and early 1960s.[86] The Second Vatican Council and Catholic critiques of colonialism led to further differences between the regime and the Church.[87]

The response to Brasil's *A Questão Sexual* was anything but moderate or neutral, and any limited Catholic enlightenment with respect to social and scientific issues – as evidenced in work by figures such as the French Jesuit Teilhard de Chardin (1881–1955) – was not a characteristic of Portuguese Catholicism in the early 1930s. Jaime Brasil's 'response to a campaign' by the Catholic *Novidades*, entitled *Os Padres e 'A Questão Sexual'*, ran to just less than a hundred pages and reproduced verbatim the original articles published by *Novidades* as well as some that appeared in other newspapers, either in favour or against Brasil's ideas. Early on in the volume, Brasil stated that his *A Questão Sexual* had been published in June 1932, was reviewed by some newspapers of the national press, such as the *Diário da Noite* and *O Século*, and was therefore known to the 'fathers' – the 'padres' – of *Novidades* very soon after publication.[88]

It was only in mid July 1932 that the 'general offensive' against *A Questão Sexual* began. Brasil was careful to point out that, apart from being a personal offensive meted out by the Catholic Church against him, it was also part of a broader attempt to counter the anti-clerical press, such as the newspaper *O Século*, for which Brasil wrote, and to admonish other recently published works that were critical of religion.[89] In response to an article by Brasil in *O Século* in favour of educational and scientific means to combat prostitution in Oporto over and above simple Catholic moral declarations, thus coinciding with the energetic hygiene campaigns of the Portuguese League for Social Prophylaxis (LPPS) based in the city, *Novidades* aimed to refute Brasil's position and published an article on 'Communism – How its propaganda is being spread in Portugal'.[90] According to this piece, Brasil, the author of the book 'of declared communist propaganda', *A Questão Sexual*, was in the process of providing almost unnoticeable homeopathic doses of 'communist venom' to be swallowed by an unsuspecting public in order to promote the social revolution he envisaged as being necessary to resolve the sexual question; this was, however, little more than 'pseudo-intellectualised pornography'.[91] Instead of chastity, marriage, honour and propriety – values classed as 'medieval' by Brasil – the scientific and moral values propounded by him were nothing more than 'the exaltation of all carnal tendencies and appetites of the human animal'.[92] In a tug of war that lasted several days, Brasil's 'intelligent and worthy' opinions were, nevertheless, defended in turn by the *Diário Liberal* the day afterwards. The author of this defence of Brasil supported his proposed solutions to the sexual question and condemned the morality of Catholicism that gave vent to 'bestial instincts'.[93]

A further article with the title 'Communist propaganda – a manual of public corruption' was printed the same day by *Novidades*, and the review reaffirmed its position, rejected Brasil's *A Questão Sexual* and declared it to be inspired by the likes of communists Alexandra Kollontai and V. I. Lenin.[94] The work, the piece ran, '[w]ishes to corrupt, above all, the youth', differentiating his work from that of Egas Moniz: 'Mr Jaime Brasil has inaugurated political pornography', which hid under 'the mask of hygiene'.[95] The book 'assaulted the most elementary precepts of morality' and pretended to justify all crimes of morality and activities 'against nature'.[96] *Novidades* went further in its acerbic critique and declared that '[t]he book by Mr Jaime Brasil aims to corrupt systematically and overtly, combatting all the defences of public morality. It is, therefore, frankly and openly revolutionary'.[97] It was, in sum, 'the most shameful manual of political corruption to come off any Portuguese press to date'.[98]

In turn, Brasil responded in the pages of the *Diário Liberal* on 21 July, refuting his supposed communist affiliation, pointing out the personal risks that such an accusation could entail and demanding that the Catholic paper provide the relevant proofs. In certain sections of his response that were not printed by the liberal paper, but which were provided in italics in his *Os Padres*, he went on to accuse figures of the Church of various misdemeanours including child molestation.[99] This merited a further response from *Novidades* and a counter-denunciation of the attitudes of the Catholic paper by Ribeiro de Carvalho (1880–1942) in the progressive *República* on 25 July, in which it was stated that Portugal was not a branch of Rome, was not subject to the 'stinking hypocrisy' of the Jesuits and was a nation free of 'ultramontane' prejudices.[100] Following this piece, Brasil elaborated on the question of Russian communism and the supposed corruption of the youth. After these exchanges, some of the same and other newspapers came out in defence of and in solidarity with Brasil. Towards the end of the year, in light of a campaign to honour him through the presentation of a golden pen for his contribution to journalism (an offering that he stated he would refuse), right-wing papers added to the condemnation of Brasil, in particular in ideological terms.[101]

The association between Brasil's work and communism, the accusation that his work contributed to the corruption of youth and the argument that scientific morality was a lesser guide for the complexities of human choice and life were all wrapped up in the *Novidades* critique. As such, the ideas that such a critique contained, if not typical of particular conservative Catholic approaches, were at least not exceptional in the Iberian setting, in which matters related to sexuality were deemed to be either sinful or essentially part of the Church's purview. The condemnation focused on the ideological aspects of Brasil's work and on the supposed corrupting elements that it proffered. Such condemnation was more general than specific, however, and particular issues were not referred to in depth. In fact, in respect of eugenics, *Novidades* did not comment on Brasil's ideas. Some comments were, however, made on the issue of contraception, and Brasil sallied forth to defend this posture by drawing on the work of Auguste Forel.[102] The vehemence with which the Catholic paper *Novidades* condemned *A Questão Sexual*, nevertheless, was extreme and, even though eugenics was not singled out for attention, the paper effectively refuted any 'scientific' meddling in an area – sexuality and reproduction – that was to be firmly maintained within the Catholic sphere.

Conclusion

The greater propensity of some Catholic intellectuals, some fifty years after Vatican II, 'to interact in a constructive way with secular reason and with science' is a relatively new phenomenon.[103] The conflict over Jaime Brasil's *A Questão Sexual* in the 1930s clearly belonged to a different era that was much more hostile to secular and scientific debates, particularly when these attempted the reform of sexual customs. The dispute between Brasil and the Jesuit *Novidades* illuminates a number of matters relevant to the history of sexuality, eugenics and religion, not only in Portugal but also internationally.

First of all, the sociopolitical circumstances of Brasil's works played a significant role in generating the raised stakes and vituperative prose that the conflict engendered. The atmosphere of 1930s Portugal, as the dictatorship of Salazar became consolidated, was one of huge political and social tensions, with 'internal enemies' on both the Left and the Right being neutralised by the regime. The struggles of the anarchist movement, whether in respect of its more trade union activities or in cultural terms, in respect of education and the sexual question, still presented a challenge to regime stability and were therefore subject to severe repression including torture, incarceration and exile. Brasil remarked in his defence against *Novidades* that the identification of someone as a communist could result in imprisonment and, in some countries, death. The fact that he was indeed imprisoned from 1940 to 1942 testifies to the reality of that threat. His work also shows that within anarchist thought on sexuality commonalities were being forged between neo-Malthusianism and eugenics; both were framed by a broad commitment to sex reform and cultural struggle, which, in turn, was anti-clerical in nature.

Secondly, the conflict between Brasil and *Novidades* is testimony to the struggle for legitimacy and the right to speak about sexual matters in Portugal. Brasil was not the only sex reformer to fall foul of the Church's condemnation; Almerindo Lessa (1909–95), the author of the slim but informative volume on eugenics *Exortações eugénicas*, was also vilified in 1933. Even the more ostensibly scientific work of someone of the stature of Egas Moniz, author of *A Vida Sexual*, did not achieve complete freedom of expression in 1930s and 1940s Portugal, his work being blacklisted by the regime and available only with a medical prescription.[104] Brasil, from a contestatory journalistic and libertarian position, was even less likely to achieve this limited freedom of expression.

Thirdly, even though eugenics was a minor issue for Catholicism in the 1930s in Portugal, and despite there being no specific reference to eugenics in the condemnation meted out to Brasil by *Novidades*, Catholics, and *Novidades* itself, had been concerned about the 'excesses' of eugenics, and far-left and far-right ideologies in Portugal and abroad since the early 1930s. While Nazism and fascism were condemned, particular opprobrium was reserved by the Catholic Church and regime for 'communism' or any Left political position. Trenchant criticism of certain forms of eugenics, and particularly sterilisation, would appear in *Novidades* as the 1930s wore on. When opposition to political extremism was combined with censoring conscious procreation, birth control and the advocacy of 'sexual revolution' it is no surprise that Brasil's 'corrupting manual' would be firmly placed in the *Novidades* firing line.

Notes

[1] This article draws on research undertaken as part of an AHRC Research Fellowship, 'Anarchism and Eugenics: A Seeming Paradox (1890–1940)' (AH/M005291/1). I am very

grateful for the support offered by the Council for this project. I am also very grateful to António Fernando Cascais for sharing some of Jaime Brasil's publications with me.

2 Jaime Brasil, *A Questão Sexual* (Lisbon, 1932). On Brasil, see João Freire and Maria Alexandre Lousada (eds), *Greve de Ventres! Para a história do movimento neomalthusiano em Portugal: em favor de um autocontrolo da natalidade* (Lisbon, 2012), p. 188, in which it is noted that he was born in Angra do Heroísmo (Azores) in 1896 and died in Lisbon in 1966. He wrote for the syndicalist paper *A Batalha* and critical cultural reviews such as *O Diabo* and *O Globo*. During the Spanish Civil War he worked in Paris, writing for the anti-fascist journals *Unir* and *Liberdade*, returning to Portugal in 1940 to be imprisoned until 1942.

3 Under the rubric of 'morbid sexuality', Brasil discussed masturbation, homosexuality, prostitution and sexual diseases. Within 'normal sexuality' he included matters relating to love, marriage, birth control and sexual freedom. Brasil stated (*A Questão Sexual*, p. 10) that only one 'serious book' on sexuality had been published in Portugal, and this was Egas Moniz's *A Vida Sexual*, which ran to multiple editions in the early twentieth century.

4 Brasil, *A Questão Sexual*, pp. 383–434, and 428–434, respectively. All translations from the Portuguese are my own.

5 Jaime Brasil, *Os Padres e 'A Questão Sexual'. Resposta a uma campanha do jornal católico 'Novidades'* (Lisbon, 1932), p. 21. It was, in fact, not the only incident of this type. The sex reformer and eugenicist Almerindo Lessa published his *Exortações Eugénicas. Notas para um programa de política genética* (Oporto, 1933) a year later. The book proved to be a scandal, and the author remembers being called 'ugly names' by members of the Catholic Church. See Almerindo Lessa, *No Tempo do meu Espaço. No Espaço do meu Tempo* (Lisbon, 1995), p. 43.

6 Brasil, *Os Padres*, p. 7.

7 See, for example, M. Turda and A. Gillette, *Latin Eugenics in Comparative Perspective* (London, 2014); G. J. Baker, 'Christianity and Eugenics: The Place of Religion in the British Eugenics Education Society and the American Eugenics Society, c. 1907–40', *Social History of Medicine*, 27/2 (2014), 281–302; G. Vallejo and M. Miranda 'Iglesia católica y eugenesia latina: un constructo teórico para el control social (Argentina, 1924–58)', *Asclepio*, 66/2 (2014), 1–12. For debates in the historical period itself, see Association du Mariage Chrétien, *L'Église et L'Eugénisme, La Famille à la Croisée des Chemins* (Paris, 1930); and J. Mayer, 'Eugenics in Roman Catholic Literature', *Eugenics, a Journal of Race Betterment*, 3/2 (1930), 43–51.

8 See D. Redvaldsen 'The Eugenics Society's Outreach to the Labour Movement in Britain, 1907–45', *Labour History Review*, 78/3 (2013), 301–29; D. Redvaldsen 'Eugenics, Socialism and Artificial Insemination: the Public Career of Herbert Brewer', *Historical Research*, 88/239 (2015), 138–60; on socialism and eugenics in France, see P.-A. Taguieff, 'Eugénisme ou décadence? L'exception française', *Éthnologie Française*, 29 (1994), 81–103; on Germany, see M. Schwartz, *Sozialistische Eugenik. Eugenische Sozialtechnologien in Debatten und Politik der deutschen Sozialdemokratie 1890–1933* (Bonn, 1995); on Soviet Russia, see A. Etkind 'Beyond Eugenics: the Forgotten Scandal of Hybridizing Humans and Apes', *Studies in History and Philosophy of Science. Part C: Studies in History and Philosophy of Biological and Biomedical Sciences*, 39/2 (2008), 205–10, and P. Simpson 'Bolshevism and "Sexual Revolution": Visualizing New Soviet Woman as the Eugenic Ideal', in Fae Brauer and Anthea Callen (eds), *Art, Sex and Eugenics: Corpus Delecti* (Aldershot and Burlington, VT, 2008), pp. 209–38; for the social democratic Left in Scandinavia, see G. Broberg and N. Roll-Hansen, *Eugenics and the Welfare State: Sterilization Policy in Denmark, Sweden, Norway, and Finland* (East Lansing, 1996).

9. Such engagement was typical of the most 'advanced' sectors of the labour movement across Europe. An impressive account of the British case is J. Rose, *The Intellectual Life of the British Working Classes* (Yale and London, 2002).
10. On the eugenics movement in Portugal, see I. F. Pimentel 'O aperfeiçoamento da raça. A Eugenia na primeira metade do século XX', *História*, 3 (1998), 18–27; A. Leonor Pereira, 'Eugenia em Portugal?', *Revista de História de Ideais*, 20 (1999), 531–600; and Richard Cleminson, *Catholicism, Race and Empire: Eugenics in Portugal, 1900–50* (New York and Budapest, 2014).
11. Freire and Lousada (eds), *Greve de Ventres!*, pp. 190–8 discuss the interest in neo-Malthusianism from the late nineteenth century up to the clandestine 'Despertar' anarchist group, which was devoted to sex education throughout the 1940s. The group distributed leaflets urging women not to procreate, or to reproduce only under certain conditions, and distributed pessaries as birth-control measures. On Portuguese anarchism in the 1940s, see João Freire, 'Os anarquistas portugueses na conjuntura do após-guerra', in Various Authors, *O Estado Novo das Origens ao Fim da Autarcia (1926–1959)*, vol. II (Lisbon, 1987), pp. 9–26.
12. See E. Rodrigues, *História do Movimento Anarquista em Portugal* (Florianópolis, 1999) and J. Freire, *Freedom Fighters: Anarchist Intellectuals, Workers, and Soldiers in Portugal's History*, trans. Maria Sousa (Montreal, 2001). For a communist-inclined history of the 18 January movement, see L. H. Afonso Manta (ed.), *O 18 de Janeiro de 1934. Do movimento de resistência proletária à ofensiva fascista* (Lisbon, 1975); and for an anarchist perspective, see J. Francisco, *Páginas do historial cegetista* (Lisbon, 1983). Anarchists were the authors of an assassination attempt on Salazar in 1937. See E. Santana, *História de um atentado: O atentado a Salazar* (Mem Martins, 1976).
13. See Á. Girón, '¿Hacer tabla rasa de la historia?: La analogía entre herencia fisiológica y memoria en el anarquismo español (1870–1914)', *Asclepio*, 52/2 (2000), 99–118; Á. Girón, *En la mesa con Darwin. Evolución y revolución en el movimiento libertario en España (1869–1914)* (Madrid, 2005). For France, see R. D. Sonn, *Sex, Violence, and the Avant-Garde: Anarchism in Interwar France* (University Park, PA, 2010).
14. R. Cleminson, *Anarchism, Science and Sex: Eugenics in Eastern Spain, 1900–1937* (Oxford and Bern, 2000). For an earlier account, see R. Á. Peláez, 'Eugenesia y darwinismo social en el pensamiento anarquista', in B. Hofmann, P. Joan i Tous and M. Tietz (eds), *El anarquismo español y sus tradiciones culturales* (Frankfurt am Main/Madrid, 1995), pp. 29–40. This relationship was played down by E. Masjuan in *La ecología humana en el anarquismo ibérico: urbanismo 'orgánico' o ecológico, neomaltusianismo y naturismo social* (Barcelona, 2000) but was broadly reaffirmed in I. Jiménez-Lucena and J. Molero-Mesa, 'Good birth and Good living. The (de)Medicalizing Key to Sexual reform in the Anarchist Media of Inter-War Spain', *International Journal of Iberian Studies*, 24/3 (2012), 219–41.
15. J. Freire and M. A. Lousada, 'O neomalthusianismo na propaganda libertária', *Análise Social*, 18/72-73-74 (1982), 1367–97; D. Duarte, 'Everyday Forms of Utopia: Anarchism and Neo-Malthusianism in Portugal in the Early Twentieth Century', in F. Bethencourt (ed.), *Utopia in Portugal, Brazil and Lusophone African Countries* (Oxford, 2015), pp. 251–73.
16. For the most extensive discussion of Darwin and the history of evolutionary theory in Portugal, see A. L. Pereira, *Darwin em Portugal. Filosofia. História. Engenharia Social (1865–1914)* (Coimbra, 2001).
17. A. P. de Lima, *A evolução do transformismo* (Oporto, 1912); C. Almaça, 'Neo-Lamarckism in Portugal', *Asclepio*, 2 (2000), 85–98; for this reception, see Cleminson, *Catholicism*, pp. 42–3.
18. The appendix is entitled 'Estudo do mendelismo, especialmente nas suas aplicações ao homem': P. de Lima, *A evolução*, pp. 117–35.

[19] P. de Lima, *A evolução*, p. 119.
[20] On Vaz, see Pereira, *Darwin em Portugal*, pp. 436–55; Freire and Lousada (eds), *Greve de Ventres!*, pp. 51–68, which reproduces some sections of Vaz, *Néo-Malthusianismo: Tese inaugural apresentada à Escola Medico-Cirurgica do Porto* (Oporto, 1902).
[21] On Paul Robin, see G. Giroud, *Paul Robin* (Paris, 1937).
[22] [João Evangelista] C. Lima, *O Movimento Operario em Portugal* (Lisbon, 1910). See Pereira, *Darwin em Portugal*, pp. 436–76. Campos Lima, as he was known, wrote the text originally in 1904 as a dissertation at Coimbra University.
[23] Lima, *O Movimento Operario em Portugal*, p. 19.
[24] C. Lima, *A Theoria Libertária ou o Anarquismo* (Lisbon, 1926), pp. 28–31. This was the printed version of a talk originally given at the Popular University.
[25] Lima, *A Theoria Libertária*, p. 29.
[26] Lima, *A Theoria Libertária*, p. 32.
[27] This would be different from the 'biocracy' of eugenicist Édouard Toulouse and first elaborated upon extensively in Portugal by L. A. Duarte Santos, 'O normotipo do homem na zona de Coimbra e o normotipo dos portugueses', *Arquivo de Anatomia e Antropologia*, VXXI (1940–1), 507–40. For Toulouse, see A. Drouard, *L'eugénisme en questions. L'exemple de l'eugénisme «français»* (Paris, 1999), pp. 21–65.
[28] J. Brasil, 'Problemas actuais. A "procriação voluntária" em nome dos superiores interesses da espécie', *O Globo. Hebdomadário de cultura, doutrina e informação*, 18 (1930), 2.
[29] Brasil, 'Problemas actuais', 2.
[30] J. Brasil, *A Procriação Voluntária. Processos para evitar a gravidez* (Lisbon, 1933). This volume was the first offering in the 'Biblioteca de Educação Sexual'. It was published by Nunes de Carvalho, the firm that also produced Brasil's *A Questão Sexual* and *Os Padres e 'A Questão Sexual'*.
[31] J. Brasil, *O Problema Sexual* (Lisbon, 1931). Like Lima's talk on anarchist theory in 1926, this was also the text of a talk given at the Popular University in Coimbra (26 March 1931); A.A. Mendes Correia, *O problema eugénico em Portugal* (Oporto, 1927).
[32] Brasil, *O Problema Sexual*, p. 8.
[33] Brasil, *O Problema Sexual*, p. 11.
[34] Brasil, *O Problema Sexual*, p. 16. Brasil does not give the full reference, but the event referred to was probably that organised in 1930 by the Association du Mariage Chrétien.
[35] Brasil, *O Problema Sexual*, p. 19 (the original Portuguese was 'Instituto de Sexualismo e Eugenia').
[36] Brasil, *O Problema Sexual*, p. 20.
[37] Brasil, *O Problema Sexual*, p. 41. In his *A Procriação Voluntária*, p. 47, Brasil refers to the attempt by the Oporto-based scholar and sex reformer Almerindo Lessa to establish a Portuguese chapter of the WLSR. Lessa wrote a piece for the Spanish League's journal in 1933. See A. Lessa 'Problemas de Psicología Sexual. El Amor', *Sexus*, 2/2 (1933), 61–71.
[38] Brasil, *O Problema Sexual*, p. 43.
[39] Brasil, *O Problema Sexual*, p. 47.
[40] Brasil, *O Problema Sexual*, p. 48.
[41] Brasil, *O Problema Sexual*, pp. 48–51. The Humberts enjoyed a prominent profile in the anarchist neo-Malthusian movement. In 1934, Jeanne Humbert was arrested and convicted of spreading neo-Malthusian propaganda in contravention of the French Law of 1920. See W. H. Schneider, *Quality and Quantity: The Quest for Biological Regeneration in Twentieth-Century France* (Cambridge, 1990), p. 185.

42 Brasil, *O Problema Sexual*, p. 51. Brasil did not state which conference, but this would have been the one convened in 1930.
43 Brasil, *O Problema Sexual*, p. 57.
44 Brasil, *O Problema Sexual*, p. 58.
45 Brasil, *O Problema Sexual*, p. 59.
46 See the bibliography in Brasil, *O Problema Sexual*, pp. 61–3.
47 See the section of Brasil, *A Questão Sexual*, pp. 428–34 on the 'selection of the species'.
48 Brasil, *A Questão Sexual*, p. 428.
49 Brasil, *A Questão Sexual*, p. 429.
50 From a different political perspective, L. Darwin, *What is Eugenics?* (New York, 1932), p. 26 argued that those filling well-paid positions should keep their levels of offspring high to maintain such positions while those in poorly paid jobs should reduce their numbers, resulting in fewer applicants for such labour, in turn meaning that wages would go up. Further, 'If the unemployed had few children, this would in like manner lessen unemployment in the future, with all its attendant misery'. As we will see below, Brasil drew on Darwin to justify his vision of eugenics.
51 Cleminson, *Catholicism*, pp. 34–40.
52 Brasil, *A Questão Sexual*, p. 429.
53 Brasil, *A Questão Sexual*, pp. 429–30.
54 Brasil, *A Questão Sexual*, p. 430.
55 Brasil, *A Questão Sexual*, pp. 430–31, where Richet's *Sélection humaine* (Paris, 1919) is quoted extensively. Richet's work is discussed in Schneider, *Quality and Quantity*, pp. 109–15.
56 Brasil, *A Questão Sexual*, p. 432.
57 Brasil, *A Questão Sexual*, pp. 469–74.
58 Brasil, *A Questão Sexual*, p. 470.
59 Brasil, *A Procriação Voluntária*, p. 7. The preface notes that the work was completed in February 1933 (Brasil, *A Procriação Voluntária*, p. 10). An eloquent example of the fusion between neo-Malthusianism and eugenics was the re-naming of Albert Gros's review, *Le Malthusien*, to become *Le Malthusien: Revue eugéniste*. See Schneider, *Quality and Quantity*, p. 37.
60 L. Huerta, 'El Malthusianismo no es el Eugenismo', *Estudios*, 77 (1930), 36–43. Huerta's book, *Natalidad Controlada*, published by the anarchist publishing house Cuadernos de Cultura (Valencia) in 1933, restates and amplifies this argument.
61 Brasil, *A Procriação Voluntária*, pp. 14–15.
62 Brasil, *A Procriação Voluntária*, pp. 17 and 47, respectively. Kehl gave his talk in Oporto on 24 October 1932. See Cleminson, *Catholicism*, pp. 106–7.
63 Brasil, *A Procriação Voluntária*, p. 19.
64 Brasil, *A Procriação Voluntária*, p. 19.
65 Brasil, *A Procriação Voluntária*, p. 20.
66 Brasil, *A Procriação Voluntária*, p. 21.
67 M. C. Stopes, *Contraception (Birth Control)* (London, 1924).
68 Brasil, *A Procriação Voluntária*, pp. 21–2.
69 Brasil, *A Procriação Voluntária*, pp. 22–3. Brasil mentioned the work by Henry Guillou, *Essai de philosophie générale élémentaire* (Paris, 1921) in this regard.
70 The original French edition was M. Devaldès, *La Maternité consciente. Le Rôle des Femmes dans l'amélioration de la race* (Paris, 1927).
71 Brasil, *A Procriação Voluntária*, p. 5, where Devaldés's (written with an acute accent in

Portugal and Spain) book is qualified as a 'lucid work'. For Devaldés's comments on the percentage formula, see Brasil, *A Procriação Voluntária*, p. 23.

[72] Turda and Gillette, *Latin Eugenics*.

[73] Brasil, *A Procriação Voluntária*, p. 25. The original Portuguese for 'reproduction' was 'proliferação'.

[74] Brasil, *A Procriação Voluntária*, p. 26.

[75] Brasil, *A Procriação Voluntária*, p. 26.

[76] Brasil, *A Procriação Voluntária*, pp. 119-21.

[77] See Cleminson, *Anarchism*, pp. 212-8 for the anarchist response to the Nazi sterilisation laws.

[78] M. B. da Cruz 'As origens da democracia cristã em Portugal e o salazarismo (I)', *Análise Social*, 14/54 (1978), 265-78; 'As origens da democracia cristã em Portugal e o salazarismo (II)', *Análise Social*, 14/55 (1978), 525-607; M. B. da Cruz, 'As elites católicas nos primórdios do salazarismo', *Análise Social*, 27/116-7 (1992), 547-74. For a succinct overview of the Church's role under Salazar, see T. Gallagher, *Portugal: A Twentieth-century Interpretation* (Manchester, 1983), pp. 125-9.

[79] A. C. Pinto and M. I. Rezola, 'Political Catholicism, Crisis of Democracy and Salazar's New State in Portugal', in M. Feldman and M. Turda (eds), *Clerical Fascism in Interwar Europe* (London/New York, 2008), p. 153.

[80] C. Pinto and I. Rezola, 'Political Catholicism', p. 154. For the changing face and evolving relationship with the regime of organised political Catholicism, the Acção Católica Portuguesa, see A. M. Ferreira, 'A Acção Católica. Questões em torno da organização e da autonomía da acção da Igreja Católica', in *O Estado Novo das Origens ao Fim da Autarcia (1926-1959)*, vol. II (Lisbon, 1987), pp. 281-302.

[81] For the early period of the journal, see R. Santos, 'O Jornalismo na Transição do Século XIX para o XX. O Caso do Diário *Novidades* (1885-1913)', *Media e Jornalismo*, 9/9 (2006), 89-104.

[82] For evidence and discussion of this position, see Cruz, 'As elites católicas', pp. 552-4, who cites several *Novidades* issues from 1931; Pinto and Rezola, 'Political Catholicism', pp. 147-8.

[83] Cruz, 'As elites católicas', p. 555.

[84] Cruz, 'As elites católicas', p. 559.

[85] P. Blanshard, *Freedom and Catholic Power in Spain and Portugal: An American Interpretation* (Boston, 1962), p. 219.

[86] N. Medeiros, 'Action, Reaction and Protest by Publishers in 1960s Portugal: Books and Other Publications in the Catholic Opposition', *Politics, Religion and Ideology*, 16/2-3 (2015), 137-53. Some translated works on conjugal love and related matters began to appear in the 1950s.

[87] R. Robinson, *Contemporary Portugal: A History* (London, 1979), p. 79.

[88] Brasil remarks (*Os Padres*, p. 11, note 1) that his use of the term 'padres' is 'employed in the most pejorative of senses'.

[89] Brasil, *Os Padres*, p. 12. Although Brasil does not state which, these may have included the above-mentioned work by Lessa.

[90] Brasil, *Os Padres*, pp. 182-0. The article was published on 18 July 1932.

[91] Brasil, *Os Padres*, p. 18.

[92] Brasil, *Os Padres*, p. 19.

[93] The article in the *Diário Liberal* is reproduced in Brasil, *Os Padres*, pp. 21-2.

[94] Brasil, *Os Padres*, pp. 22-5. Later, *Novidades* referred to the book as 'verdadeiramente

95 anarquizadora' ('truly anarchistic'), adding to the lack of clear ideological classification. See Brasil, *Os Padres*, 'Em duas palavras', *Novidades* (31 July 1932), 67.
95 Both quotes from Brasil, *Os Padres*, p. 23.
96 This must have referred to Brasil's approval of female sexual freedom and contraception rather than of homosexuality and masturbation, which he decried.
97 Brasil, *Os Padres*, p. 25.
98 Brasil, *Os Padres*, p. 25.
99 Brasil, *Os Padres*, pp. 27–30. He also referred to the immorality of Catholics in their frequenting of houses of prostitution, of unsavoury practices in convents and of 'pederasty' (p. 50); the latter charge is repeated (p. 56).
100 Brasil, *Os Padres*, p. 37.
101 See, for example, Brasil, *Os Padres*, pp. 78–9.
102 Brasil, *Os Padres*, p. 56. An article in the *Diário de Notícias*, 30 July 1932, with the title 'A Questão Sexual por Jaime Brasil', rated Brasil's work alongside that of Egas Moniz and Forel, in Brasil, *Os Padres*, p. 67.
103 L. Oviedo and A. Garre, 'The Interaction between Religion and Science in Southern Catholic Europe (Italy, Spain, Portugal)', *Zygon*, 50/1 (2015), 173.
104 *Comissão do Livro Negro sobre o Regime Fascista: Livros proibidos no regime fascista* (Sintra/Mem Martins, 1981), p. 67.

RESPONSIBLE PARENTHOOD: REPRODUCTION AND RELIGION IN POST-WAR BRITAIN

Patrick T. Merricks

'Man's development is influenced, not only by inborn qualities and dispositions and by environment, but also by a spiritual factor lying beyond both – namely, the grace of God.'[1] Thus wrote Reverend Derrick Sherwin Bailey (1910–84) in *The Eugenics Review* in 1959. In the 1950s and 1960s, British churchmen and eugenicists shared two major interests: the growing population, and emerging fertility treatments. In engaging with these issues, this article explores a subject of increasing interest and importance to present and future society. The introduction outlines issues of agreement and contention between eugenicists and churchmen on 'responsible parenthood'; the first section details debates on contraception surrounding the 1958 Lambeth Conference and the passing of new government legislation on birth control and abortion a decade later; the second section explores contrasting opinions on artificial insemination and its use to improve society; and the conclusion draws parallels between these opinions and how parenthood is discussed today. If eugenic and religious opinions were influential in shaping the understanding of reproductive technologies and habits in mid-century Britain, implementing new scientific techniques to improve the human species should heighten our inherent sense of moral responsibility.

Introduction

After the Second World War, the Anglican Church carefully monitored the world's growing population. Its concerns were articulated in publications such as Richard Fagley's *The Population Explosion and Christian Responsibility*.[2] In 1960, the Professor of Christian Ethics Victor Obenhaus (1904–94) bemoaned that 'Intercontinental ballistic missiles (ICBM), [the] reunification of Germany, nuclear fall-out, and the stored surpluses of agriculture and industry [...] have obscured the view of what may in the long run transcend all these other issues, namely the population explosion'.³ The heightening tensions of the Cold War distracted many from the long-term problem of global population growth; Anglicans were concerned not only with mankind's spiritual destiny, but with its future as a species.

At the 1958 Lambeth Conference, the Anglican Church expressed its desire to play down the traditional Christian role of reproduction as the sole purpose of marriage. Instead, a loving marriage was one that made responsible use of contraception – a moral choice that benefitted the family and the community. The Anglican Church had gone from rejecting birth control at the 1920 Lambeth Conference, to a reluctant acceptance for married couples in 1930, to presenting it in 1958 as the key to a happy marriage and healthy society. In a passage that highlights the overlaps between eugenic and religious views, the Archbishop of Canterbury, Geoffrey Fisher (1887–1972) was quoted in a newspaper report at the time thus: 'Responsible parenthood, built on obedience to all the duties of marriage, requires a wise stewardship of the resources and abilities of the varying population needs and problems of society and the claims of future generations.'⁴ These concerns were shared by politicians and culminated in the government's passing of the 1967 National Health Service (Family Planning) and Abortion Acts. To the delight of eugenicists and churchmen, the former 'made contraception readily available through the NHS by enabling local health authorities to provide advice to a much wider population'.⁵ The latter was the beginning of the end for dangerous and illegal 'backstreet' abortions in the UK and, as a last resort, a means safely to avoid unwanted children being born into an inflated population. The contraceptive pill (readily available after the Family Planning Act) became both a symbol of and a physical means to achieve female independence and emancipation from the 'traditional' role of wife and mother.

To eugenicists, the spread of birth control among the highly fertile, and usually poorer sections of society or 'bad stocks/problem families',⁶ was one of the last tools available for controlling the biological 'quality' of the population at a time when eugenics was heavily discredited. National eugenic societies had influenced public policy and the nature of everyday life in the Americas, Europe and Asia. The involvement of German racial hygiene with Nazism was catastrophic for eugenic movements around the world. In Britain, the movement failed to regain the following that it had held before the war. *The Eugenics Review* ceased publishing in 1968, and eventually the Eugenics Society became the Galton Institute in 1989. Tellingly, one of the most outspoken eugenicists after the Second World War was a prominent churchman: Bishop E. W. Barnes of Birmingham (1874–1953). Until his death, Barnes warned of the dangers of 'racial intermixture' and pushed for the implementation of sterilisation and infant euthanasia for mental defectives in the new welfare state. Barnes was, however, an anomaly within the Eugenics Society and the Anglican Church, with few in either institution agreeing with his scientific racism or negative eugenics. Even so, recent scholarship on Barnes has revealed how polarised Britain was on the issue of eugenics after the Second World War, with many surprisingly in favour of preventing the 'inhuman' from reproducing.[7]

In the 1950s and 1960s, the Eugenics Society believed that the philosophy of responsible parenthood could finally address its chief concern throughout the twentieth century: the differential birth rate among the classes. If rarely by name, eugenicists still influenced British social policy with their continued involvement in associations such as Planned Parenthood, originally founded by the Eugenics Society in 1930 as the National Birth Control Council.[8] The legacy of the British eugenics movement survives today in the form of global charities such as the International Planned Parenthood Association and Marie Stopes International. On a smaller scale, the Eugenics Society also helped found UK charities such as the Family Planning Association and Brook Advisory Centres. With the help of the Internet and social media, such institutions now offer a range of contraceptive procedures and advice on sexual health to an ever-increasing number of young, vulnerable adults.

The Anglican Church published numerous studies on emerging reproductive treatments, including 1962's *Human Reproduction: A study of some emergent problems and questions in the light of the Christian Faith*. The 45-page report dealt with contraception, artificial insemination, abortion and sterilisation. It was praised in *The Eugenics Review* for its 'straightforwardness and clarity of purpose' and highlighted the 'advance in Christian thought over the last few decades and the stage in its evolution so far reached'.[9] However, if there was agreement on contraception, this was not the case for artificial insemination (AI), now widely identified as in vitro fertilisation (IVF) – the most successful technique to achieve artificial insemination. In the 1940s, the Archbishop of Canterbury organised an advisory group to establish the

Anglican Church's position on the practice, where it was denounced as wholly unnatural and immoral. During the 1960s, as the technology developed many theologians began to argue that 'AIH' (using the 'H'usband's sperm) was acceptable for married couples who could not conceive 'naturally'. However, 'AID' (from a 'D'onor) was deplored by churchmen and perceived as being fit for only a Huxley (either Aldous or Julian)-inspired dystopian future. In contrast, AID offered eugenicists the technology needed for national genetic improvement: the problem was convincing a population all too aware of the Nazi experiment that eugenics could have any positive influence at all.

Birth control

Popular attitudes to responsible parenthood have interested a range of scholars since the Second World War. Richard Hoggart's multi-edition book *The Uses of Literacy* (1957) provides a detailed critique of the habits of the working class. He observed that there were 'few working-class areas in which a substantial proportion of people still attend church or chapel', yet they remained 'in some sense a part of the life of the neighbourhood'.[10] Aside from attending Sunday school as children, many in Britain did not frequent church every Sunday; yet, most still attended religious institutions at 'the important moments of life or in times of personal crisis' such as marriage ceremonies, baptisms and funerals. As Hoggart commented, 'they are not simply taking out a saving policy; they still believe underneath, in certain ways'.[11] Elsewhere, authors such as Nigel Yates and Callum Brown agree that secularisation, the sexual revolution, and the short-lived religious revival of the early 1950s have been overstated in their immediate impact on the general population.[12] For example, the sexual revolution (the spread of contraceptive practices and rise of recreational sex) was felt mostly by the middle and upper classes, leaving the working class relatively untouched into the 1970s.

In post-war Britain, some Anglican Churchmen continued to sympathise with eugenics in its 'purest' form: the furtherance of human evolution. Eugenics and religion were not necessarily incompatible. It was a theologian, Dean Inge (1860–1954), who was among the first to draw popular attention to eugenics in the 1910s and 1920s, and after 1945 Bishop Barnes still had his supporters within the Anglican Church, albeit in greatly diminished numbers. The Eugenics Society arguably spent more time discussing birth control than any other eugenic measure; from 1909 until 1968 the phrase was used in *The Eugenics Review* twice as many times as 'sterilisation'. In 1956, the former Society General Secretary and a leader of Planned Parenthood, Carlos Blacker (1895–1975) claimed that responsible parenthood brought 'the eugenics and birth control movements together; and unlike the standards based on class, it excites no prejudices'.[13] Influenced by a lengthy memorandum submitted by the Eugenics Society at the tail end of the Second World War,[14] the 1949 Royal Commission on Population drew national attention to the importance of birth control, concluding that: 'The giving of advice on contraception to married persons who want it should be accepted as a duty of the National Health Service and the existing restrictions on the giving of such advice by public authority clinics should be removed.'[15] Despite this and the efforts of eugenicists and the birth control movement in establishing advice centres across the country, it was almost two decades before responsible parenthood was given its legal backing with the 1967 Family Planning Act.

The Anglican Church began to re-evaluate its official position on birth control with these issues in mind. Moreover, it is no coincidence that the Church's growing acceptance of birth control since 1930 coincided with the vast availability and comparative sophistication of contraception in mid-century Britain. Writing for the eugenic Population Investigation Committee on 'Birth Control and the Christian Churches', Flann Campbell (1919–94) claimed that 'a simple policy of silence or disapproval is not enough. Social realities must be faced, arguments met and answered, new formulae invented'.[16] World population has grown exponentially since the nineteenth century, which for many Anglicans justified a deviation from the traditional Christian approach; birth control had fast become a moral necessity in civilised society. For the most part, this was not the case for Catholics, who were bound to papal authority. In 1958, Pope Pius XII (1876–1958), although equally interested in emergent reproductive technologies, declared that any form of contraception was a 'grave violation of moral law' that interfered with the sacred act of procreation – something that public authority had no right, under any pretext, to permit.[17] Although most churchgoers in Britain were Protestant, a significant minority were Catholic and were guided by their local religious officials in line with the teachings of the Vatican. One observer commented that should the Anglican Church simply uphold its tentative acceptance of birth control as established in 1930 at Lambeth, 'the English bishops will have displayed an agility which a few Roman cardinals could almost bring themselves to envy'.[18]

Further reference to Hoggart reveals a negative contemporary view of the working class's breeding habits that is shared by many academics: 'as in most aspects of domestic life, [...] the wife is expected to be responsible for contraceptive practice'. Having received limited guidance prior to marriage on family planning and the like, this was 'a degree of sustained competence many wives are hardly capable of. She forgets just once or "lets herself go", or a sheath is cheap and bursts, or the husband demands awkwardly after a night at the club. How often, therefore, it is assumed that any children after the first were "not intended"'.[19] Within the middle classes the rare occurrence of usually a third child, 'who was not intended" is apt to arrive when the parents are about forty', some ten years after the first two. In contrast, unplanned working-class children were 'likely to arrive only a year or two after the others. It is usually accepted "philosophically"; after all "what did yer get married for?"'.[20] Until the 1958 Lambeth Conference, the Anglican Church, too, held that the sole purpose of marriage was to bring children into the world, regardless of the negative impact that this had on the family and community. Blacker described the Church's official position up to this point: 'The resolutions of the Lambeth Conference of 1908 and 1920 denounced contraception outright; that of 1930, carried by a three to one majority, was grudgingly permissive.'[21] Such findings reinforced the eugenicist's belief that the philosophy of responsible parenthood was yet to touch poor communities.

In early 1958, a report, entitled 'The Family in Contemporary Society', was published at the behest of Archbishop Geoffrey Fisher of Canterbury by the Church of England's aptly named Moral Welfare Council. If eugenicists influenced the Anglican Church's opinions on population and family life to any degree, this is confirmed by the presence of Derrick Bailey, a member of the Eugenics Society, as the Council's 'Study Secretary'. Bailey was an outspoken figure, having authored *Homosexuality and the Western Christian Tradition* in 1955, which paved the way for the 1957 Wolfenden Report and the eventual decriminalisation of homosexuality. In publishing 'The Family in Contemporary Society', he intended to sway the Anglican Bishops meeting in the summer months at Lambeth Palace towards an official pronouncement in favour of the widespread use of contraception for family limitation. The Anglican Church was becoming more progressive in some respects; yet, the Council also called women's education 'the most disruptive force in all matters of family life'.[22] This echoed arguments by British eugenicists during the Second World War, which had led to a lengthy and polarised debate on the role of women in society between male and female contributors to *The Eugenics Review*. Could a wife achieve fulfilment in the workplace and the academic sphere, while also raising a family and satisfying the needs of her husband? Both the Anglican Church and the Eugenics Society wished to preserve traditional gender roles – a key reason that they could unite under the philosophy of responsible parenthood.

One contributor to the above-mentioned debate was Barbara Bosanquet (1906–87), who declared that, despite its misgivings, the 1958 Report proved that 'the Anglican Church is concerned with family and population problems on a realistic basis. Also, it narrows the gulf between reverent agnostics and believers'.[23] A shared focus on the family, between churchman and eugenicist, blurred the line between Christianity and humanism that appeared so pronounced to many observers of (and contributors to) mid-twentieth century postmodern thought. Agreeing with the sentiments of the Council, Barbara Bosanquet argued that: 'The best family (from a Christian and not merely a humanist point of view) is not that in which there are most children but that in which children grow up in an atmosphere of confidence and affection, fed by the "mutual endearment" of parents who engage in regular intercourse without thereby producing more children than they can manage to rear.'[24]

The proceedings of the Council made national news. In what it called 'the most striking part', *The Manchester Guardian* summarised the publication's key argument thus: 'The classical Christian attitude [disapproval] was formulated at a time when natural checks like war and disease seemed to guarantee the world against over-population. The advance of medical science has transformed the moral problem.' Universal healthcare was a dream long realised in Britain after the 1948 creation of the National Health Service. That infant mortality rates were at an all-time low owed equal thanks to improvements in public health, advances in medicine, understanding of diet, training of doctors, medical equipment and so on, during the first half of the twentieth century and before. The report argued that this profound moral concern and sense of urgency should be transferred to spreading the philosophy of responsible parenthood: '"Family Planning" represents an extension of the responsible use of science into the realm of procreation, in the immediate interest of the family and the more remote but no less real interest of society.'[25] Never in Britain had Christians and eugenicists agreed to such an extent.

The 1958 Lambeth Conference was important for national acceptance of birth control and abortion. In terms of the latter, the attendees pled for 'compassionate understanding of the predicaments which have compelled families and governments in some parts of the world to regard abortion as a more merciful thing than starvation'.[26] Indeed, *The Times* reported that 'the acute problems of population in parts of the world like India and the Far East had caused the ninth Lambeth Conference to deal with family planning'.[27] It was hoped that Britain could set an example for the rest of the world in responsible parenthood. Derrick Bailey called its conclusions 'rightly critical of the tradition which for so long has taught that procreation is the role or the principle purpose of marriage'.[28] Resolutions 112–31 were dedicated to 'The Family in Contemporary Society' under the premise that 'all problems of sex relations, the procreation of children, and the organisation of family life must be related, consciously and directly, to the creative, redemptive, and sanctifying power of God'.[29]

The nineteenth-century ideal that society could gradually progress towards higher states of enlightenment had been eroded by a range of factors – not least modernisation, the mass movement of populations, war, economic collapse and extreme ideologies. All, in one way or another, seemed to threaten the traditional Christian idea of a healthy family. The Anglican Church also bemoaned other aspects of modern life, including the corrupting influences of journalism, radio and television – 'those systems of migratory labour that break up family life' – and, worse still, drugs and alcohol, or, rather, the 'widespread and growing reliance on undesirable and artificial means of responding to the restlessness of our present age'.[30] The weakening of family life and social cohesion from the 'sheer silliness and the low standard of social morality' threatened permanently to retard spiritual well-being in the Christian world. If the views of the Church reflected those of the established elite, it was in wishing to maintain public order by protecting the 'traditional' family structure. This was confirmed with the Drugs (Prevention of Misuse) Act 1964 and the Dangerous Drugs Act 1965, which, in accordance with various decrees from the United Nations, criminalised the cultivation and use of cannabis and amphetamines, among other substances, in Britain.[31]

The 1958 Lambeth Conference also discussed family planning at length. Resolution 115 expressed that 'the responsibility for deciding upon the number and frequency of children has been laid by God upon the consciences of parents everywhere; that this planning, in such ways as are mutually acceptable to husband and wife in Christian conscience, is a right and important factor in Christian family life'. No longer was 'sexual love an end in itself nor a means to self-gratification' but rather 'self-discipline and restraint [. . .] essential conditions of the freedom of marriage and family planning'.[32] Bailey expressed his delight to the Eugenics Society that 'the unity and harmony between the different functions or aspects of sexuality can be broken more easily than ever'. While the Old Testament encouraged people to be fruitful and multiply, it was written at a time when 'under-population rather than over-population was the dominant reality'.[33] In accepting the social value of responsible parenthood, the Church now committed itself to adapting its teachings to support the family in what the Archbishop called 'an age of rapid social change'.[34] In a total inversion of 'traditional' religious responses to birth control, the 1958 Lambeth Conference recognised churchmen as the leaders in spreading knowledge on responsible parenthood, welcoming 'the increasing care given by the clergy to preparation for marriage both in instructing youth, through confirmation classes and other means'.[35] This was the responsibility not just of the clergy but of lay people, too: 'The Conference welcomes the growth of marriage guidance councils, [. . .] given as a Christian vocation by well-trained Christian husbands and wives, is a volunteer service of great value, makes an important contribution to the community, and deserves government support.'[36]

Bailey also discussed the eugenic significance of the Conference at a Members Meeting for the Eugenics Society in November 1958. The bishops spent considerable time discussing eugenic sterilisation – some even sympathising with the practice before eventually dismissing it. While 'its value as a family planning device suited to poor and illiterate peoples was appreciated, though not approved', it was incompatible with the Christian idea of a free person.[37] Nevertheless, with churchmen taking an active interest in limiting the number of children produced within marriage, Bailey concluded that eugenicists should be 'profoundly grateful to the bishops for their labour on "The Family in Contemporary Society", and we may confidently expect to see it bear fruit before their lordships are again invited to confer at Lambeth in ten years' time'.[38]

Bailey's prophecy proved accurate, as the push for government support was successful with the passing of the 1967 National Health Service (Family Planning) and Abortion Acts. In the intervening years, the Eugenics Society also went to great efforts to promote responsible parenthood. A key figure here was Helen Brook (1907–97). In 1958, thanks to her admirable volunteer work for Carlos Blacker's Planned Parenthood Association (she also went on to serve as its chair and then president in the 1960s and 1970s), the Eugenics Society entrusted Brook to take over Marie Stopes's birth control clinic after Stopes's death. According to Lesley Hall, unlike the Planned Parenthood Association and marriage guidance offered by the Anglican Church, Stopes's clinic did not confine itself to married or soon-to-be married women alone; it focused more on poor, single women who sought advice on contraception at a time when there were no official bodies to turn to. In 1964, she established the Brook Advisory Centres, which offered strictly confidential counsel on reproduction to all young, single people – men and women – under the age of 25. By the end of the 1960s, the centres had expanded nationwide, offering advice to over 10,000 unmarried people, with most being aged between 19 and 25 and a sixth under 19.[39] Half a century later, alongside the website 'brook.org.uk', the charity 'Brook' helps 250,000 young people across the country.[40] Considering the centres' humanist (and Christian) ideals that provide aid to such people in desperate situations, it is curious (and perhaps somewhat jarring) that they were originally created by the Eugenics Society to reduce swelling numbers in the working class.

The National Health Service (Family Planning) Act 1967 confirmed that *ir*responsible parenthood was a national issue. To combat this, the Act made contraception readily available through the NHS by 'enabling local health authorities to provide advice to a much wider population. Previously, these services were limited to women whose health was put at risk by pregnancy'. The MP Edwin Brook had put it forward as a private member's bill, having identified 'a social problem whereby low income groups were at risk of economic struggle through having more children than they could afford'.[41] The Act also protected the confidentiality of those seeking advice. For example, a 1971 issue of the *British Medical Journal Supplement* reveals a case in which a family doctor had received a letter from a Brook Advisory Centre to prescribe the contraceptive pill to a girl of 16. After informing the girl's parents, the doctor was examined and eventually cleared of misconduct for breaching confidentiality.

Providing a glimpse into the mindset of the public, the mother expressed her relief thus: 'Well, if it has to be, thank God she had enough sense to protect herself.'[42]

In the same year as the Family Planning Act, the government also passed the Abortion Act. Despite its prior illegal status, abortion also had a prominent place in the national psyche. Alongside contraception, it featured in films such as *Alfie* (1966), starring Michael Caine and Shelley Winters, works of fiction such as Penelope Mortimer's *Daddy's Gone A-Hunting* (1958) and non-fiction such as Storm Jameson's *Journey to the North* (1969) and *The Nameless: Abortion in Britain Today* (1966), all of which discuss and document at length the social stigmas surrounding the practice and different means to end pregnancy outside the law.[43] The Abortion Act 1967 made the procedure acceptable, albeit only in a medical emergency. The strictness of the Act is evidenced by several cases brought to the General Medical Council's Disciplinary Committee. For example, a 'Dr Mansoor Soomro' was charged with professional misconduct and imprisoned for aborting the child of a 16-year-old after 'The pregnancy was confirmed but nothing was said about there being any danger or risk of injury to the girl if it was allowed to continue'.[44]

Although the Abortion Act 1967 legalised the procedure in the case of medical emergencies, the 'pro-choice' abortions that we know today (at least in Britain) were not an option; life-threatening 'backstreet' abortions and home-brewed noxious concoctions provided the only solution for poor and desperate women. Instead, unwanted pregnancies continued to produce unwanted children. It was not until 1990 that these laws were relaxed and abortions became more freely available after the passing of the Human Fertilisation and Embryology Act. Notably, one new stipulation that allowed for abortion after the 1990 Act was that if 'there is a substantial risk that if the child were born it would suffer from such physical or mental abnormalities as to be seriously handicapped'. The 1990 amendments have allowed more people from poorer backgrounds to undergo abortion and, in some cases, prevent children from being born with what eugenicists may have termed 'mental deficiency'. While not the case in 1967, the Act has evolved to a state in which it could be used for eugenic purposes if popular opinion were swayed in favour of the eugenic cause (in this case, responsible parenthood and ridding the world of hereditary disease). The government archive page concludes that the 1967 Family Planning and Abortion Acts meant that 'women were able to take control of their own fertility for the first time'.[45] While the Acts are remembered as part of the women's liberation cause, the Family Planning Act was passed (at least partially) in the interests of controlling the population and its genetic progress, and the Abortion Act for only exceptional circumstances in which the procedure could save the mother's life.

Artificial insemination

In post-war Britain, many began thinking seriously about the uncertain future of the human species. Certainly, the possibility of nuclear war increased throughout the 1950s, and almost became a reality due to events in Berlin and Cuba at beginning of the 1960s. The nuclear arms race was a threat to humanity's evolution, whether through the potentially degenerative effects of radiation on human genes or, more broadly, the destruction of the planet. Many turned to artificial insemination and sperm banks to protect the species. In 1958, Carlos Blacker claimed that 'if the threat of all-out nuclear war is sufficient to require the continuous maintenance in the air of American planes loaded with H bombs, it should also call for the construction of underground seminal banks, protected from radiations'.[46]

Artificial insemination was nothing new in mid-twentieth-century Britain. The physician John Hunter (1728–93) performed the first documented application in eighteenth-century London. The technique was developed slowly, and in the nineteenth and early twentieth centuries there took place experiments conducted by figures such as J. Marion Simms in the United States, Walter Heape at Cambridge and V. K. Milovanov in Russia. New effective methods to preserve sperm, and a greater understanding of the menstrual cycle, meant that the technique could also be applied to livestock with great success, beginning in the inter-war United States.[47] By 1960, it was used for all forms of UK livestock apart from chicken.[48] Eugenicists agreed that its use in agriculture had 'much shortened the mental step involved in recognising its bearings on human infertility and even, potentially, on human improvement'.[49] The first successful human pregnancy using frozen sperm also took place in the United States in 1953. However, this technique was highly controversial at the time and was described as 'contrary to public policy and good morals'.[50] Ombelet and Robays have suggested that 'it is not surprising that nearly a decade passed before the first successful birth from frozen sperm was announced in public: a major breakthrough in history'.[51] Artificial insemination brought with it a host of moral issues. It was generally agreed, along with Derrick Bailey, that the procedure 'allows generation to be abstracted completely from any kind of personal relation'.[52] This had profound and surprising effects on the way in which the practice was eventually accepted in society. In 1958, divorce courts in Edinburgh and Chicago ruled in the wives' favour that AID was not adultery, as it did not involve sexual intercourse.[53] *The Manchester Guardian* suggested that now that most objections had been undermined by the law, it would give doctors more confidence in performing the procedure.[54]

British eugenicists took significant interest in artificial insemination, and it featured in over a hundred *Eugenics Review* articles. AID represented the most rational and efficient means of selective breeding. While negative eugenics – including contraception, abortion and sterilisation – dealt largely with the problems of the present, artificial insemination was focused directly on the future. While Marie Stopes and J. B. S. Haldane had flirted with the idea in the early twentieth century, it first became a significant eugenic discourse in 1935 with Herbert Brewer.[55] In his article 'Eutelegenesis', Brewer claimed that it would 'transform the problem of negative eugenics'. Whereas 'the elimination of [...] degeneracy by sterilizing' would be like 'clearing a river of fish by catching the few which jump from the water', AID would ensure that the 'existence of the whole inextricable tangle of latent defect' would be swept out 'in a few generations, replacing it concurrently with hereditary material of the highest excellence'.[56]

Together with Julian Huxley (1887–1975) and Hermann J. Muller (1890–1967), Brewer was part of a circle of eugenic ideologues (which also included the likes of Haldane and George Bernard Shaw) who favoured the political Left and saw artificial insemination as the future of human reproduction. As Huxley had put it: 'the inherent diversity and inequality of man is a basic biological fact; and Eugenics is the expression of a wish to utilize that fact in the best interests of future generations'.[57] In 1937, Brewer expanded on this: 'the aims of eutelegenesis are socialism, biological socialism [...] nothing less than socialization of the germ plasm, the establishment of the right of every individual that is born to the inheritance of the finest hereditary endowment that anywhere exists'.[58] The proposal was met with much caution, with his overt references to socialism doing nothing to quell any opposition. Other concerns ranged from fears about public reaction to practical difficulties such as 'finding [an] acceptable method of obtaining semen'.[59]

It would be long into the post-war period before these issues were overcome. Even then, in the first couple of decades the service was used exclusively by married couples. As Martin Richards summarised: 'A couple's desire to have their "own" children has always seemed stronger than any eugenic aspirations.'[60] Even so, in 1950 Blacker described how AID had been successfully used for eugenic purposes 'in cases when the male partner is at fault and also when, because of hereditary infirmities in himself or his family, he does not want children of his own'. However, the position of the Eugenics Society was that 'this revolutionary biological innovation [. . .] should not be legally prohibited'.[61] Later, Bailey also referenced isolated cases in which couples had used AID to avoid passing on the husband's epilepsy.

After the Nazi experiments of selective breeding, no one articulated more caution towards artificial insemination than the Anglican Church. The sexual relationship between a man and wife was sacred; yet, artificial insemination conjured images of Aldous Huxley's *Brave New World*, in which babies are grown outside the body and raised communally in a godless society. Moreover, in 1949, at a debate in the House of Lords, the Archbishop of Canterbury, Geoffrey Fisher, called AID adultery.[62] A Methodist Conference in 1958 agreed that the practice was 'a breach of the marriage vow'. The same year, the Lambeth bishops argued that this 'would not be in any way diminished by resolving the legal difficulties which at present surround the practice'.[63] At the time, objections to artificial insemination were not moral alone. In the days before IVF, when fertilisation still took place in the womb, there were significant health risks for the mother. This was enough of an issue for the British Medical Association to advise against the use of AID at its 1950 AGM.[64]

In 1945, the Church of England formed a committee to discuss artificial insemination. Members included Bishop Barnes of Birmingham and the Bishops of Derby (Alfred Rawlinson) and Oxford (Kenneth Kirk). The group felt an initial repugnance to the idea, unable to overcome the seemingly 'unnatural' nature of the practice. The three bishops then interviewed, among others, the Minister of Health, several prominent doctors and Carlos Blacker. Rather than condemn the practice outright, the best course of action for AIH 'as between husband and wife', was simply that 'judgement should be reserved'.[65] A eugenicist himself, Barnes was torn on the issue, confessing to the Archbishop of Canterbury that:

If the Church yielded to the very strong instinct which moves most of us personally, [. . .] it might be that in years to come men [would] point to our decision as another example of the way in which, when a new departure was made, the Church sought to block the way of progress.[66]

For AID, it was concluded that 'grave exception should be taken from the Christian point of view'.[67]

With these issues considered, in 1959 the British Council of Churches formed an Advisory Group on Sex, Marriage and the Family. In addition to its ecclesiastical members, the Council sought the advice of secular authorities. Derrick Bailey was a key member of the Advisory Group, as was the Medical Secretary of the Family Planning Association, Eleanor Matthews, and the Secretary of the Marriage Guidance Council, A. J. Bradshaw. The Group's findings were published in 1963 as *Human Reproduction: A Study of Some Emergent Problems and Questions in the Light of the Christian Faith*. The study had two aims: first, to inform both clergy and laity about recent and possible future developments in human reproduction; and, second, to discuss some basic questions raised by these techniques. On contraception, the authors agreed with the position established at Lambeth in 1958. Indeed, among the foremost of mankind's problems were 'the perils to peace and to the whole quality of human life of an uncontrolled birth-rate', and birth control was seen as being a moral necessity. However, there was now unparalleled power to control population size, and it was said that this should be met with caution: 'How much is man called upon to be an active and deciding fellow-worker with God? [. . .] Is this a power which invades the sphere which should be reserved for God's sole action?'[68] With birth control accepted, it was asked where the line should be drawn on interfering with procreation. This question pervades the whole publication. Moreover, while eugenics is not wholly denounced, the connection between selective breeding and disastrous philosophies such as German National Socialism is reiterated throughout as a stark warning to all'.

Influenced by Bailey, Matthews and Bradshaw, the report displays reluctant sympathy for eugenics. Although it is recommended that family planning should guide 'those contemplating marriage or parenthood who suffer from diseases obviously capable of being inherited', sterilisation was not accepted because in the most severe cases of mental deficiency the patient was unable to give consent. The Group was less certain about isolated cases. Indeed, Bailey admitted that AID had been used in families in which the father had epilepsy: 'which, of course, is a eugenic use of this technique'. Notably, the committee made it clear that 'eugenic considerations have their place' in national planning, suggesting, for instance, 'tax arrangements' for 'those who would benefit the community most by reproducing their kind'.[69] With negative eugenics greatly criticised at the time, many post-war eugenicists focused almost exclusively on positive eugenics such as this. There is also evidence to suggest that universities paid lecturers to have more children, also for eugenic purposes. The founder of the welfare state, William Beveridge (1879–1963), admitted to the Eugenics Society during the war that the London school of Economics had adopted this. As Julian Huxley noted: 'negative eugenics is of minor evolutionary importance and the need for it will gradually be superseded by efficient measures of positive eugenics'.[70]

Human Reproduction features a lengthy discussion on artificial insemination and reproductive technologies. The Advisory Group tentatively accepted AIH if used to 'bring a child into an infertile marriage'.[71] However, the reader is warned that in the field of human reproduction 'the greatest care and judgement' must be exercised 'in evaluating the contribution which scientific techniques may bring to a truly human life'. This view was universal at the time. Richard Hoggart has documented that most working-class families believed that for all of the developments by 'the scientist', from penicillin and atomic energy to contraception and artificial insemination, 'working-class people persist in assuming that there is a straight-forward moral responsibility in both the act of engagement in, and the application of, such discoveries'.[72] With contraception, mankind could control 'the number of persons on this earth', but with artificial insemination 'also the kind of persons they shall be'. It would not be long before 'we shall be confronted with far-reaching eugenic proposals' in which we can 'choose the sex of our children. This is a simple choice, but massively charged with potential effect on the future'. The comparison is made with the Nazi Lebensborn project, in which 'superior' Nazi officers would father children with multiple Aryan mothers, to be raised communally. This was recognised by eugenicists. Blacker considered the idea that while eminent men and women could 'be courteously asked by the President of a National Eugenics Corporation to contribute to posterity a portion of their sexual glands', to some 'these possibilities will appear fascinating, to others abominable'.[73] Thus, it was asked: should such techniques be used to 'enable the nation to breed from the best stocks?'

To answer this question, the committee considered the ideal Christian family. God's purpose for procreation was that 'persons should enter this world through the closest of all relationships, whereby two persons become "one flesh" and that, by the semen of the one fertilising the ovum of the other, another person should be conceived possessing genetic characteristics of the two persons thus fused together'. It was doubted that a national eugenic programme of selective breeding could ever be reconciled with this modern Christian conception of a free person. Moreover, while intelligence, physical strength and the presence of hereditary disease could be subject to genetic control, the committee was 'unaware of any claim by responsible eugenicists to be able to breed spiritual or moral capacities'.[74] Eugenics could not be used to bring mankind closer to God. This could be achieved only by preserving the freedom to choose your life partner – this is what drove the evolution of mankind, as per the conclusions of the Advisory Group. After careful consideration, AID was rejected. It both alienated mankind from God and could be abused by totalitarian governments to breed humans like livestock (the technique had only recently been applied to cattle breeding). According to Barbara Bosanquet, the use of AID 'on a large scale for purely eugenic purposes would be wholly unjustifiable from the Christian standpoint'.[75]

While the Eugenics Society gave its support to AID and the Anglican Church did not, it has never been used in a national eugenic programme. Despite the reservations of the Anglican Church, sixteen years later, in 1978, the first 'test tube baby', Louise Brown, was born in the UK. The development of IVF brought artificial insemination into the mainstream. After the Human Fertilisation and Embryology Act 1990, it became available in Britain from a donor or otherwise on the NHS and privately, with waiting lists long and costs high respectively. Meanwhile, the 'sperm bank' became a commercial industry in the United States and, more recently, has reached mainstream cultural status, with humorous portrayals in episodes of well-known series such as *Friends* (1996, 2003) and *Parks and Recreation* (2013). Notably, the storylines for both shows feature independent women with successful careers wishing to become pregnant and raise children alone. This both goes against the attempts of mid-twentieth-century British churchmen and eugenicists to preserve traditional family structure and confirms their fear that artificial insemination could divorce procreation from marital union.

Conclusion

In the 1950s and 1960s, British eugenicists and churchmen successfully campaigned for the introduction of new laws on birth control and abortion. The Lambeth Conference of 1958 remains a landmark moment for the Anglican Church, solidifying its status as a progressive Christian institution that kept the moral interests of the community at its forefront. The Eugenics Society helped to push through legislation on family planning to reduce the working-class birth rate. Either way, contraception and abortion are now more readily available than ever in Britain. Eugenicists also believed that AID could be used to ensure that future generations would have no genetic disease. However, the Anglican Church was generally repulsed by the idea: the sacred act of procreation was between man and wife only. AID has been developing ever since, and is used by more people than ever, whether by independent women, same-gender couples or, as eugenicists and churchmen wished, in the 'traditional' family of man and wife.

Contrasting opinions on artificial insemination also continue today. 'Be special, give sperm' is the advertising slogan of Britain's largest sperm bank, located in London. However, not everyone is suitable for this elite group of special sperm givers. In 2015, *The Guardian* published an article on the 'eugenic' nature of the London Sperm Bank, which described 'a policy of turning away autistic donors and those diagnosed with other neurological disabilities, such as attention deficit hyperactivity disorder [ADHD], dyslexia and obsessive compulsive disorder'. On behalf of the Autistic Self Advocacy Network, Ari Ne'eman argued that 'eliminating autism, dyslexia and other similar disabilities might remove valuable talents, [and] such changes may leave humanity less equal, less diverse, and perhaps even less human'.[76] This is comparable to the approach of Anglican churchmen who, in the 1960s, argued that humanity's variation was part of God's evolutionary plan. Were the Anglicans right in saying that variation is an essential component of humanity's success?

Elsewhere, the term 'designer babies' has been in vogue for some time. Early this year, the development of three-person babies (a technique that uses donor mitochondria to prevent inheritable mitochondrial diseases such as deafness and epilepsy) has also led leading scientists such as Madhumita Murgia to suggest that 'Eugenics is a dirty word, most commonly associated with racist profiling, or Nazi experiments. But the time has come to rethink our attitude'.[77] While isolated cases may not fulfil the eugenic goal of controlling human evolution, the fact that the term's association with gas chambers and concentration camps is beginning to wane is significant.

Finally, overpopulation remains a serious concern. Last year, China brought an end to its 35-year-old 'one child' policy. Originally designed to combat population growth, it became widely associated with forced abortions, sterilisations and even infanticide. Many have heralded the decision as a 'rare human rights victory in a country where freedoms are tightly restricted', agreeing with the Anglican warnings in *Human Reproduction* on how new techniques of reproduction could be abused in dictatorships. In response to these global fears, Bill Gates (whose father was head of Planned Parenthood) – one of the world's richest men – has put a massive proportion of his fortune into spreading the once-eugenic philosophy of responsible parenthood to the Third World. Now, more than ever, the arguments put forward by post-war eugenicists may seem valuable as we look once again to the future of the human species.

Notes

[1] D. Bailey, 'The Lambeth Conference and the Family', *The Eugenics Review*, 50/4 (1959), 245.
[2] R. Fagley, *The Population Explosion and Christian Responsibility* (New York, 1960).
[3] V. Obenhaus, 'Review: The Population Explosion and Christian Responsibility', *The Journal of Religion*, 40/4 (1960), 314.
[4] G. Fisher, quoted in 'Family Planning Guidance by the Bishops', *The Times* (26 August 1958), 3.
[5] 'National Health Service (Family Planning) Act 1967', *Parliamentary Archives* (1967) HL/PO/PU/1/1967/c39 (*www.parliament.uk/about/living-heritage/transformingsociety/ private-lives/relationships/collections1/parliament-and-the-1960s/national-health-service-family-planning-act/* (accessed 12 September 2016)).
[6] See Carlos Blacker, *Problem Families: Five Inquiries* (London, 1952).
[7] See P. T. Merricks, 'Checking a "Scrub" Population: Bishop Barnes of Birmingham on Inner City Slums', in Marius Turda (ed.), *Revista de Antropologie Urbana*, 5 (2015), 41–53 and P. T. Merricks, '"God and the Gene": E.W. Barnes on Eugenics and Religion', *Politics, Religion and Ideology*, 13/3 (2012), 253–74.
[8] See M. Pyke, 'Family Planning: An Assessment', *The Eugenics Review*, 55/2 (1963), 71–9.
[9] 'Notes of the Quarter', *The Eugenics Review*, 54/4 (1963), 188.
[10] R. Hoggart, *The Uses of Literacy* (Harmondsworth, 1971), 112 [originally published 1957].
[11] Hoggart, *The Uses of Literacy*, p. 114.
[12] See N. Yates, *Love Now, Pay Later? Sex and Religion in the Fifties and Sixties* (London, 2010) and C. G. Brown, *Religion and Society in Twentieth-Century Britain* (London, 2006).
[13] C. Blacker, 'Family Planning and Eugenic Movements', *The Eugenics Review*, 47/4 (1956), 231.

14. Author Unknown, 'Royal Commission on Population: Memorandum submitted by the Eugenics Society', *The Eugenics Review*, 37/3 (1945), 92–104.
15. 'A Royal Commission', *The Times* (17 April 1937), 3.
16. F. Campbell, 'Birth Control and the Christian Churches', *Population Studies*, 14/2 (1960), 143.
17. 'Pope Pius XII: One of His Last Talks to Doctors', *The British Medical Journal*, 2/5102 (18 October 1958), 970.
18. B. Bosanquet, 'The Family in Contemporary Society', *The Eugenics Review*, 50/2 (1958), 127.
19. Hoggart, *The Uses of Literacy*, p. 45.
20. Hoggart, *The Uses of Literacy*, p. 46.
21. C. Blacker quoted in Bosanquet, 'The Family in Contemporary Society', p. 127.
22. Bosanquet, 'The Family in Contemporary Society', p. 126.
23. Bosanquet, 'The Family in Contemporary Society', p. 126.
24. Bosanquet, 'The Family in Contemporary Society', p. 126.
25. 'Church and Family' and 'The Family in Contemporary Society', *The Manchester Guardian* (15 April 1958), 3–8.
26. Bailey, 'The Lambeth Conference and the Family', p. 245.
27. 'Family Planning Guidance by the Bishops', *The Times* (26 August 1958), 3.
28. D. Bailey, 'The Lambeth Conference and the Family', *The Eugenics Review*, 50/4 (1959), 240.
29. 'Resolution 112', in *The Lambeth Conference: Resolution Archive from 1958*, 28 (*www.anglicancommunion.org/media/127740/1958.pdf* (accessed 27 September 2016)).
30. 'Resolution 127', in *The Lambeth Conference*, p. 32.
31. 'Drugs (Prevention of Misuse) Act 1964', *Parliamentary Archives* 64 (HL/PO/PU/1/1964/c36 accessed 4 October 2016).
32. 'Resolution 113', in *Parliamentary Archives*, p. 29.
33. Fagley, *The Population Explosion and Christian Responsibility*, p. 15.
34. G. Fisher, quoted in 'Notes of the Quarter', *The Eugenics Review*, 50/3 (1958), 156.
35. 'Resolution 114', in *Parliamentary Archives*, p. 29.
36. 'Resolution 117', in *Parliamentary Archives*, p. 29.
37. Bailey, 'The Lambeth Conference and the Family', p. 244.
38. Bailey, 'The Lambeth Conference and the Family', p. 244.
39. L. Hall, 'Half a Century of Helpful People with Helpful Answers', *From the Collections* (12 July 2014), Wellcome Library (*http://blog.wellcomelibrary.org/2014/07/half-a-century-of-helpful-people-with-helpful-answers/#* (accessed 4/10/2016)).
40. By the time of her death in 1997, there were 18 branches in the UK, funded by local health authorities, and, today, advice is freely available on the website in the form of videos and FAQs. See: 'Brook, the young people's sexual health & wellbeing charity' (*https://www.brook.org.uk/* (accessed 4 October 2016)).
41. 'National Health Service (Family Planning) Act 1967' in *Parliament and the 1960s* (*https://www.parliament.uk/about/living-heritage/transformingsociety/private-lives/relationships/collections1/parliament-and-the-1960s/national-health-service-family-planning-act/* (accessed 4 October 2016)).
42. Author Unknown, 'Professional Secrecy', *British Medical Journal Supplement*, 1/5750, (20 March 1971), 79.
43. See B. Brookes, *Abortion in England, 1900–1967* (London, 1988); and L. Hall, *Literary Abortion* (*http://www.lesleyahall.net/abortion.htm* (accessed 26 October 2016)). Although this is an underdeveloped area of research, work is currently being conducted on 'Representations of Abortion in British Literature and Film from around 1918' at the

44. Author Unknown, 'Imprisonment for Abortion', *British Medical Journal Supplement*, 1/5750 (20 March 1971), 77–8.
45. 'Medical Termination of Pregnancy', in *Abortion Act 1967 [1990 amendments]* (http://www.legislation.gov.uk/ukpga/1967/87/section/1 (accessed 4 October 2016)).
46. C. Blacker, 'Artificial Insemination: The Society's Position', *The Eugenics Review*, 50/1 (1958), 51.
47. W. Ombelet and J. Van Robays, 'Artificial Insemination History: Hurdles and Milestones', *Facts Views Vis Obgyn*, 7/2 (2015), 140.
48. 'Control of Artificial Insemination', *The Times* (9 June 1960), 14.
49. Blacker, 'Artificial Insemination', p. 51.
50. Ombelet and Robays, 'Artificial Insemination', p. 140.
51. Ombelet and Robays, 'Artificial Insemination', p. 141.
52. Bailey, 'The Lambeth Conference and the Family', p. 241.
53. See 'Insemination not Adultery', *The Times* (11 January 1958), 6; 'Legality of Artificial Insemination: Chicago Judge's Ruling', *The Times* (14 December 1954), 8; 'Artificial Insemination not Adultery', *The Manchester Guardian* (11 January 1958), 5.
54. 'Effect of Lord Wheatley's ruling', *The Manchester Guardian* (13 January 1958), 5.
55. See M. Richards, 'Artificial Insemination and Eugenics: Celibate Motherhood, Eutelegenesis and Germinal Choice', *Studies in History and Philosophy of Biological and Biomedical Sciences*, 39/2 (2008), 211–21.
56. Brewer, 'Eutelegenesis', p. 123.
57. J. Huxley, 'Foreword', in H. Brewer, *Eugenics and Politics* (London, 1937), p. 1.
58. H. Brewer, *Eugenics and Politics* (London, 1937), p. 3.
59. Martin Richards, 'Artificial Insemination and Eugenics: Celibate Motherhood, Eutelegenesis and Germinal Choice', *Studies in History and Philosophy of Science Part C*, 39/2, (2008), 211 (213).
60. Richards, 'Artificial Insemination and Eugenics', 211.
61. Blacker, *Statement of Objects*, p. 8.
62. 'Lords Debate on Artificial Insemination', *The Times* (17 March 1949), 4.
63. Bailey, 'The Lambeth Conference and the Family', p. 244.
64. Bailey, 'The Lambeth Conference and the Family', p. 244.
65. J. Barnes, *Ahead of His Age: Bishop Barnes Of Birmingham* (London, 1979), p. 429.
66. Ernest W. Barnes, 'Reply to Fisher RE: Artificial Insemination,' (20 February 1945), EWB 9/21/10.
67. 'Committee on Artificial Insemination: Draft Interim Report' (est. July 1945), EWB 9/21/28.
68. The British Council of Churches, *Human Reproduction: A Study of Some of the Emergent Problems and Questions in the Light of Christian Faith* (London, 1962), p. 29.
69. The British Council of Churches, *Human Reproduction*, pp. 26 and 30.
70. J. Huxley, 'Eugenics in Evolutionary Practice', *The Eugenics Review*, 54/3 (1962), 135.
71. The British Council of Churches, *Human Reproduction*, p. 30.
72. Hoggart, *The Uses of Literacy*, p. 45.
73. Blacker, 'Artificial Insemination', p. 53.
74. Blacker, 'Artificial Insemination', p. 53.
75. B. Bosanquet, 'Review: *Human Reproduction*', *The Eugenics Review*, 55/1 (1963), 37.
76. Ari Ne'eman, 'Screening sperm donors for autism? As an autistic person, I know that's the road to eugenics', *The Guardian* (30 December 2015) (http://www.theguardian.com/

commentisfree/2015/dec/30/screening-sperm-donors-autism-autistic-eugenics (accessed 7 November 2016)).

[77] M. Múrgia, 'Eugenics need not be a dirty word – instead, it could be lifesaving technology', *The Telegraph* (26 October 2015) (*https://www.telegraph.co.uk/technology/news/11956083/ Eugenics-isnt-a-dirty-word.-Instead-could-be-lifesaving-technology.html* (accessed: 7 November 2016)).

INDEX

A
Aquinas, T. 42

B
Bailey, D. S. 85, 90, 92–4, 97–100
Barnes, E. W. 87–8, 98
Bateson, W. 1
Beveridge, W. 100
Blacker, C. 89–90, 94, 96, 98, 100
Bombarda, M. 69
Bosanquet, B. 91, 101
Brasil, J. 61–78
Brewster, D. 20

C
Campbell, F. 89
Castro, F. de 50–1
Carrel, A. 12
Chalmers, T. 20
Chardin, T. de. 75
Conroy, J. 21
Cooper, J. M. 39
Correia, A. A. M. 67

D
Darwin, C. 4, 17, 42, 64
Darwin, E. 17
Darwin, L. 70, 72
Daubeny, C. 19
Davenport, C. 9
Devaldès, M. 72

Dubourg, M.-L. 11
Duns, J. 20

F
Fisher, G. 86, 90, 98
Fleming, J. 20
Floch, H. Le 53–4
Fonssagrives, J. 53
Fülöp, Z. 9

G
Galton, F. 1–8, 10–13, 71
Gasparri, P. 57
Gates, B. 103
Gemelli, A. 56
Giovanni, G. de 52–4
Green, T. H. 20
Gore, C. 20
Gotti, G. 50

H
Haldane, J. B. S. 1–2, 97
Holland, H. S. 20
Hoornaert, G. 53
Hunter, J. 96
Huxley, A. 98
Huxley, J. 12, 71, 97, 99–100
Huxley, T. H. 19–20
Hürth, F. 55–6

I
Inge, W. 4, 10, 88

Index

K
Kelsen, H. 13
Kidd, B. 28–9
Kingsley, C. 22
Koltsov, N. 9
Knox, J. 19
Kuyper, A. 6

L
Lamarck, J-B. 17, 64
Lankester, E. R. 26
Laughlin, H.H. 54
Lecomte, A. J. 45–6
Ledóchowski, W. 56
Leo XIII 42
Lessa, A. 78
Levinas, E. 13
Liberatore, M. 43
Lima, A. P. 64–5
Lima, J. E. C. 65, 72

M
Madzsar, J. 9
Matthews, E. 99
Mayer, J. 39
Melcher, P. L. 50
Miller, H. 20
Mitchell, P. C. 6
Moscati, G. 52
Moody, D. 21, 25
Moore, A. 19–21, 31–5
Muckermann, H. 11, 39
Muller, H. J. 97
Mügge, M. A. 9

N
Ne'eman, A. 102
Newman, J. H. 19

O
Obenhaus, V. 85
Osborn, F. 2, 13, 71

P
Parente, A. 50–1
Parrocchi, L. M. 49
Paul VI 41
Pearson, K. 5
Pius XI 12, 39, 42, 51, 53, 55–6
Pius XII 89
Porro, E. 49–50
Pouchet, F. A. 45
Poulton, E. B. 21

R
Rainy, R. 20, 31
Richet, C. 70
Robin, P. 65
Rolleston, G. 19
Ryan, J. A. 39

S
St Athanasius 32
St Augustine 32
Saleeby, C. 9
Sankey, I. 21, 25
Slaughter, J. W. 10
Sixtus V 50
Stopes, M. 70, 87, 94, 97

T
Talbot, E. S. 20
Temple, F. 19
Thomson, W. 21

V
Val, R. M. del 56
Vaz, Â. 65
Vermeersch, A. 53, 55

W
Wallace, A. R. 17, 26–7
Wesley, J. 19
Wiggam, A. E. 2
Whitney, L. F. 2

ISBN 978-1-78683-378-5
eISBN 978-1-78683-379-2
ISSN (Print)   2057-4517
ISSN (Online) 2057-4525
The Journal of Religious History, Literature and Culture
© University of Wales Press, 2018
Articles and reviews © The Contributors, 2018

Contributors to *The Journal of Religious History, Literature and Culture* should refer enquiries to the journal page at www.uwp.co.uk or e-mail press@press.wales.ac.uk requesting notes for contributors.

Advertising enquiries should be sent to the Sales and Marketing Department at the University of Wales Press, at the address below.

Subscriptions: *The Journal of Religious History, Literature and Culture* is published twice a year in June and November. The annual subscription for institutions is £95 (print only), £85 (online only) or £140 (combined); and for individuals is £25 (print or online only) or £40 (combined). Subscription orders should be sent to University of Wales Press, University Registry, King Edward VII Avenue, Cardiff CF10 3NS. E-mail: press@press.wales.ac.uk.